Cation and Anion Chemistry

CATION AND ANION CHEMISTRY

James H. Krueger
Oregon State University

Bogden & Quigley, Inc.
Publishers

Tarrytown-on-Hudson, New York / Belmont, California

1971

Copyright © 1971 by Bogden & Quigley, Inc., Publishers.

All rights reserved. No part of this book may be reproduced, stored in a retrieval system, or transcribed, in any form or by any means, electronic, mechanical, photocopying, recording, or otherwise, without the prior written permission of the publisher, 19 North Broadway, Tarrytown-on-Hudson, New York 10591.

Cover Design by Aaron Hirsh

Text design by Science Bookcrafters, Inc.

Library of Congress Catalog Card No.: 79-164945

Standard Book No.: 0-8005-0012-1

Printed in the United States of America

2 3 4 5 6 7 8 9 10 — 75 74 73 72

Preface

The chemistry of a number of common cations and anions is surveyed in this laboratory text. Particular emphasis is given to equilibrium systems in aqueous solution and to theory and experiments concerning the nature of transition metal ions and coordination compounds. The chemical principles developed in Part I relate directly to the experimental work; therefore, this material should be covered along with the laboratory work in Parts II and III. Although Chapters 1 through 6 are self-contained, they can be used to supplement corresponding topics in a textbook of general chemistry.

A major objective of this laboratory textbook is to involve students with real chemical systems. It is important that the student become familiar with the experimental, as well as the theoretical aspects of chemistry. The reactions of common cations and anions in aqueous solution provide an interesting and easily accessible system for students to investigate. The experiments involve the analysis and identification of selected ions. This allows a thorough examination of their properties while avoiding the detail and lengthy procedures which are a part of classical analytical schemes. Although analytical aspects are not stressed, several unknowns are included because many students enjoy the challenge and continuity of working with unknown mixtures. Emphasis is placed on observing and understanding the reactions of the ions. Chapters 1 through 6 provide the general principles needed to interpret ion reactions in aqueous solution. In addition, Chapters 8 through 11 and Chapter 13 provide specific information on the chemical properties of the ions. Considerable emphasis is placed on the coordination chemistry of transition metal cations and on the importance of anions as ligands.

This laboratory textbook has been designed for flexible use in the general chemistry program. It is self-contained and can be used alone or to supplement the textbook used in the lecture. By omitting some of the unknowns, or by omitting anion chemistry in Part III, the time required can be adjusted to fit the number of laboratory periods available. Either cation chemistry in Part II or anion chemistry in Part III can be studied first. The analysis of solid salts and mixtures of salts in Chapter 14 is optional. Experiments on these unknowns provide a greater challenge since procedures are not presented in detail.

The material in this laboratory text has evolved from an earlier version, *Cation Chemistry*. The experimental procedures have been thoroughly tested over several years in general chemistry classes at Oregon State. Many students and graduate instructors have provided suggestions for improvement of the material. In particular, the efforts of Professors Darwin Reese and Max Williams in class-testing the experiments are greatly appreciated. Thanks are due to Professors Russell Drago and Daniel Huchital for reviewing the preliminary version and providing a number of helpful suggestions.

J.H.K.

Corvallis, Oregon
January 1971

Contents

Preface   *v*

PART I | General Characteristics of Cations and Anions

1   The Study of Ions in Water Solution   *1*
     1.1 Chemical Objectives   *1*
     1.2 Nature of Ions in Water   *2*
     1.3 Names and Formulas   *3*

2   Reactions of Cations and Anions   *7*
     2.1 Equilibrium in Water Solution   *7*
     2.2 Acids and Bases   *8*
     2.3 Salts of Low Solubility. Precipitation Reactions   *12*
     2.4 Oxidation–Reduction Reactions   *13*
     2.5 Formation of Complex Ions   *18*

3   Writing Chemical Equations   *21*
     3.1 Writing Correct Products and Reactants   *21*
     3.2 Balancing Chemical Equations   *23*

4   Equilibrium Calculations   *31*
     4.1 Dissociation of Weak Acids and Bases   *32*
     4.2 pH. Measuring Acidity in Solution   *35*
     4.3 Solubility of Salts   *37*
     4.4 Simultaneous Equilibria. Precipitation of Sulfides   *41*

5   Coordination Compounds of Transition Metals   *49*
     5.1 Complex Ions   *49*
     5.2 Naming Coordination Compounds   *51*
     5.3 Coordination Numbers and Structures of Complex Ions   *52*
     5.4 Isomerism   *54*

6   Properties of Complex Ions   *59*
     6.1 Bonding   *59*

6.2	Spectral and Magnetic Properties	62
6.3	Reactions of Complex Ions	64

PART II | Cation Chemistry and Procedures

7 Guide to Experimental Work — 71
- 7.1 The First Laboratory Period — 71
- 7.2 Laboratory Safety — 72
- 7.3 Reagents — 73
- 7.4 Experimental Techniques — 74
- 7.5 Using this Laboratory Textbook — 76

8 Group 1: Ag^+, Pb^{2+}, and Hg_2^{2+} — 79
- 8.1 Chemistry of Group 1 — 79
- 8.2 Group 1 Procedure — 83

9 Group 2: Hg^{2+}, As(III), Pb^{2+}, Cu^{2+}, and Cd^{2+} — 91
- 9.1 Chemistry of Group 2 — 91
- 9.2 Group 2 Procedure — 97

10 Group 3: Fe^{3+}, Ni^{2+}, Al^{3+}, and Cr^{3+} — 109
- 10.1 Chemistry of Group 3 — 109
- 10.2 Group 3 Procedure — 115

11 Groups 4 and 5: Ba^{2+}, Ca^{2+} and Na^+, NH_4^+ — 127
- 11.1 Chemistry of Groups 4 and 5 — 127
- 11.2 Groups 4 and 5 Procedure — 131

PART III | Anion Chemistry and Procedures

12 Guide to Experimental Work — 149
- 12.1 The First Laboratory Period — 149
- 12.2 Laboratory Safety — 150
- 12.3 Reagents — 150
- 12.4 Experimental Techniques — 150

13 Anion Chemistry — 151
- 13.1 General Plan for Anion Identification — 151
- 13.2 Chemistry of the Anion Test Systems — 153

13.3	Procedure for Known Anions	*155*
13.4	Procedure for Unknown Anions	*159*

14 Salts and Mixtures of Salts *173*

Appendix *187*

Index *193*

PART I | General Characteristics of Cations and Anions

The Study of Ions in Water Solution

CHAPTER 1

1.1 CHEMICAL OBJECTIVES

This laboratory textbook deals with the chemistry of some of the important cations and anions in water solution. The experiments are designed to provide an opportunity to become familiar with the chemical behavior of these ions. They involve practical situations in which you should be able to relate chemical principles to experimental observations. The experiments are structured in a way that leads to a systematic means of identifying individual cations and anions. The analytical techniques for identifying ions such as mercury(II), copper(II), sulfite, and carbonate are of practical importance. These, and other ions, play useful and, in some cases, undesirable roles in environmental chemistry. This book considers only the more common ions and the characteristic reactions that lead to their identification. Other, more rigorous textbooks of qualitative analysis are available which offer very detailed schemes for testing for the presence of a large number of ions.

Overall Plan of the Experiments

The chemistry of cations is covered in Part II. (The term **cation** refers to any positively charged species, such as Na^+, Cu^{2+}, and NH_4^+.) In working with the cations, it will be necessary to carry out separations into small groups of similar ions. Each of these groups will be analyzed further into individual ions so that characteristic test reactions can be carried out. The general plan for separating the common cations into five subgroups, Groups 1 through 5, is shown in Figure 1.1. Each of the groups is separated from the others on the basis of precipitation reactions that produce salts of low solubility in water. (The general aspects of precipitation reactions are discussed in Chapters 2 and 4.) The general cation diagram (Figure 1.1) indicates that addition of HCl to a mixture of all the cations involved results in the precipitation of the chlorides $AgCl$, $PbCl_2$, and Hg_2Cl_2. The corresponding cations, Ag^+, Pb^{2+}, and Hg_2^{2+}, constitute Group 1. Separation of these ions leaves all the remaining ions in solution. Those ions which form insoluble sulfides in acidic solution constitute Group 2. The Group 3 ions are insoluble in basic sulfide solutions, and the Group 4 ions are separated as insoluble carbonates. The soluble ions, Na^+ and NH_4^+, remain to

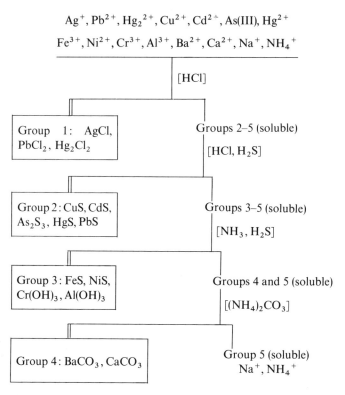

Figure 1.1 ⦚ *General Cation Diagram*

form Group 5. The chemistry of Group 1 is examined first. Preliminary experiments are carried out on a mixture known to contain each of the Group 1 cations, followed by experiments on an unknown mixture in which the identity of the ions present is to be determined. After completing work on the five groups, a general unknown, which may contain any combination of the 15 cations, is analyzed.

The chemistry of anions is covered in Part III. (The term **anion** refers to negatively charged species, such as OH^-, Cl^-, and SO_4^{2-}.) Individual anions, rather than mixtures of anions, are examined in Part III. The anions are classified into five subgroups, Groups A through E. The anions in each of the groups possess similar chemical characteristics. Identifying tests are applied, first to known anions, and then to a series of individual, unknown anions.

1.2 ⦚ NATURE OF IONS IN WATER

The chemistry of plants and animals, of rivers, lakes, and the ocean, and of many industrial processes depends on chemical reactions that occur in water solution. A large number of these reactions involve cations and anions. Thus, a knowledge of the chemical behavior of ions in water is essential.

At first, it may seem surprising that separated positive and negative ions can exist in solution. The energy released when ions combine to form ionic solids or neutral molecules is usually very large. A wide variety of ions can exist in water because of the special nature of solvent water. Liquid water has a high dielectric constant and the individual H_2O molecules are polar:

$$O^{\delta-}$$
$$^{\delta+}HH^{\delta+}$$

Ions interact strongly with water molecules in a process referred to as **hydration**. Cations are attracted to the negative oxygen portion of the water molecules and anions interact with the partially positive hydrogen atoms. Depending on their nature, ions may be surrounded by 1, or perhaps as many as 10 to 20, water molecules. In some cases the number of hydrating water molecules is well defined, as, for example, in the $Cr(H_2O)_6^{3+}$ ion. Hydration interactions can be great enough to counterbalance electrostatic attractions between cations and anions. Thus, separated ions of opposite charge can exist in water solution. However, precise physical measurements reveal that ions still exert mutual effects (ion-activity effects) on neighboring ions, even in dilute solution.

Pure liquid water is essentially a molecular substance. It is dissociated only slightly to give hydrogen ions and hydroxide ions:

$$H_2O + H_2O \rightleftarrows H_3O^+ + OH^- \tag{1}$$

The chemistry of a number of ions is modified by their ability to increase the production of hydrogen ions or of hydroxide ions in water solutions. Such ions are referred to as acids or bases, respectively, and are discussed in Chapter 2.

1.3 | NAMES AND FORMULAS

The formulas and names of a number of common cations and anions are listed in Table 1.1. This table should be useful in naming compounds and as a guide to possible products in writing chemical equations. For example, the common sulfur-containing anions are sulfate, sulfite, sulfide, and thiosulfate.

Some of the cations in Table 1.1 exhibit variable charges, or they may be said to exist in more than one oxidation state (see Section 2.4). The Roman numeral in parentheses after the name denotes the oxidation state of the ion. Note that the oxidation number is equal to the charge on a monatomic ion. For example, the two common cations Fe^{2+} and Fe^{3+} are called iron(II) and iron(III), respectively. The older, but still commonly encountered names, ferrous and ferric for Fe^{2+} and Fe^{3+}, can be used as well. Since the use of the suffixes *-ous* and *-ic* may be confusing if one is not familiar with all the oxidation states of a metal, the Roman-numeral designation will be employed in this book.

The central nonmetal element in many of the oxyanions also exists in two or more oxidation states: for example, the chlorine-containing oxyanions in Table 1.1. The commonly used names are listed in the table, but the Roman-numeral system may also be used. In this system, the chlorine oxyanions would be designated ClO^-, chlorate(I); ClO_2^-, chlorate(III); ClO_3^-, chlorate(V); and ClO_4^-, chlorate(VII). The numbers refer, not to the charge, but to the oxidation state of chlorine (see Section 2.4).

Table 1.1 | *Common Cations and Anions*

Monatomic Cations

Na^+	sodium	Fe^{2+}	iron(II) (ferrous)
K^+	potassium	Fe^{3+}	iron(III) (ferric)
Mg^{2+}	magnesium	Cu^+	copper(I) (cuprous)
Ca^{2+}	calcium	Cu^{2+}	copper(II) (cupric)
Ba^{2+}	barium	Co^{2+}	cobalt(II) (cobaltous)
Al^{3+}	aluminum	Co^{3+}	cobalt(III) (cobaltic)
Pb^{2+}	lead	Cr^{3+}	chromium(III) (chromic)
Ag^+	silver	Hg^{2+}	mercury(II) (mercuric)

Polyatomic Cations

NH_4^+	ammonium	H_3O^+	hydronium
Hg_2^{2+}	mercury(I) (mercurous)		

Monatomic Anions

F^-	fluoride	H^-	hydride
Cl^-	chloride	S^{2-}	sulfide
Br^-	bromide	O^{2-}	oxide
I^-	iodide		

Polyatomic Anions

OH^-	hydroxide	ClO^-	hypochlorite
CN^-	cyanide	ClO_2^-	chlorite
SCN^-	thiocyanate	ClO_3^-	chlorate
NO_2^-	nitrite	ClO_4^-	perchlorate
NO_3^-	nitrate	SO_3^{2-}	sulfite
CO_3^{2-}	carbonate	SO_4^{2-}	sulfate
SiO_4^{4-}	silicate (ortho)	HSO_4^-	hydrogen sulfate
PO_4^{3-}	phosphate	$S_2O_3^{2-}$	thiosulfate
AsO_4^{3-}	arsenate	CrO_4^{2-}	chromate
$C_2H_3O_2^-$	acetate	MnO_4^-	permanganate

Salts

The names of ionic compounds resulting from appropriate combinations of cations and anions can be obtained directly from Table 1.1. The cation is named first, including the Roman numeral for metals with variable oxidation states, and the anion is named second:

KBr	potassium bromide
$MgCl_2$	magnesium chloride
$Ba(MnO_4)_2$	barium permanganate
$CoCl_2$	cobalt(II) chloride
$CoCl_3$	cobalt(III) chloride
NH_4ClO_4	ammonium perchlorate

Acids and Bases

Bases and basic oxides are named by the method used for salts:

KOH	potassium hydroxide
HgO	mercury(II) oxide
$Fe(OH)_2$	iron(II) hydroxide
$Fe(OH)_3$	iron(III) hydroxide

Acids that do not contain oxygen are named using the form *hydro...ic acid*, or alternatively by using the name "hydrogen" preceding the name of the anion:

HF	hydrofluoric acid	or	hydrogen fluoride
HCl	hydrochloric acid	or	hydrogen chloride
HCN	hydrocyanic acid	or	hydrogen cyanide

Acids derived from oxyanions are named using the characteristic name of the corresponding oxyanion and the forms *-ic acid*, if the anion name ends in *-ate*, or *-ous acid*, if the anion name ends in *-ite*. (Consult a general chemistry textbook for a detailed discussion of naming these and more complex acids.)

Acid		Corresponding Sodium Salt	
H_2CrO_4	chromic acid	Na_2CrO_4	sodium chromate
H_2SO_3	sulfurous acid	Na_2SO_3	sodium sulfite
H_2SO_4	sulfuric acid	Na_2SO_4	sodium sulfate
HClO	hypochlorous acid	NaClO	sodium hypochlorite
$HClO_2$	chlorous acid	$NaClO_2$	sodium chlorite
$HClO_3$	chloric acid	$NaClO_3$	sodium chlorate
$HClO_4$	perchloric acid	$NaClO_4$	sodium perchlorate

Problems

1. Write formulas for the following compounds.

    a. Hydrobromic acid_____

    Barium hydroxide_____

    Aluminum sulfate_____

    Sodium nitrite_____

    Copper(I) iodide_____

    Copper(II) iodide_____

    Ammonium dihydrogen phosphate_____

    b. Silver oxide_____

    Ferric phosphate_____

    Cupric chromate_____

    Thiosulfuric acid_____

    Cobalt(II) phosphate_____

    Potassium aluminum sulfate_____

2. Name each of the following compounds.

    a. HNO_3_____

    HNO_2_____

    $NaHSO_4$_____

    $CoPO_4$_____

    $Hg(NO_3)_2$_____

    $Hg_2(NO_3)_2$_____

    $Ca(C_2H_3O_2)_2$_____

    $Co(OH)_3$_____

The Study of Ions in Water Solution

   b. H_2S _____

     K_2S _____

     H_3AsO_4 _____

     $KHSO_3$ _____

     $HgSO_3$ _____

     PbO _____

     NH_4MnO_4 _____

3. Write formulas for all the possible compounds that can be formed from the ions listed below. Only compounds that contain one type of cation and one type of anion should be considered.

   H^+ Mg^{2+} Al^{3+} I^- SO_4^{2-} PO_4^{3-}

Reactions of Cations and Anions

CHAPTER 2

Cations and anions undergo a variety of reactions in water solution. The most important types of reactions are discussed in this chapter: acid–base reactions, the formation of salts of low solubility, oxidation–reduction reactions, and complex-ion formation. The purpose of this chapter is to examine the general nature of these processes, to better understand the chemistry of cations and anions encountered in the experimental work. Quantitative aspects of these reactions are discussed in Chapter 4.

2.1 { EQUILIBRIUM IN WATER SOLUTION

The occurrence of a chemical reaction between two generalized species A and B can be represented as

$$a\text{A} + b\text{B} \rightarrow c\text{C} + d\text{D} \tag{1}$$

In this reaction, products C and D are produced from reactants A and B. If it is also true that C and D can react together to produce a measurable amount of A and B, the reaction is reversible. The ability to react in both directions is expressed by

$$a\text{A} + b\text{B} \rightleftarrows c\text{C} + d\text{D} \tag{2}$$

When A and B are mixed in solution, only the forward reaction, equation (1), takes place at first. As the concentrations of products C and D increase, the reverse reaction can occur also. The rates at which the two processes occur depend on the concentrations and conditions involved. After sufficient time has elapsed, both the forward and reverse reactions proceed at the same rate. When this occurs, the system is defined to be in chemical equilibrium. The equilibrium state is characterized by constant concentrations of each of the four species. No *net* change is observable in an equilibrium system. However, on the molecular level, A and B are continually being converted into C and D, and vice versa. Chemical equilibrium is a dynamic situation on the molecular level. The generalized equilibrium system in equation (2) is described mathematically by the relationship

$$K = \frac{[\text{C}]^c[\text{D}]^d}{[\text{A}]^a[\text{B}]^b}$$

K is called the equilibrium constant and has a constant numerical value at a given temperature. The larger the value of K, the greater the concentration of products and the lower the concentration of reactants at equilibrium.

As a specific example of a reversible system, consider the dissociation of acetic acid dissolved in water:

$$HC_2H_3O_2 + H_2O \rightleftarrows H_3O^+ + C_2H_3O_2^- \qquad (3)$$

Acetic acid molecules and water molecules react to produce the hydronium ion, H_3O^+, and acetate ion. If the reverse reaction did not occur as well, all the $HC_2H_3O_2$ molecules would dissociate into ions. However, the product ions have an appreciable tendency to recombine. Reaction (3) is reversible and quickly comes into chemical equilibrium. Because acid-dissociation reactions of this type are important in aqueous chemistry, they are discussed in detail in Sections 2.2 and 4.1.

A highly important feature of equilibrium systems is their tendency to shift or adjust when conditions are changed. The **principle of Le Châtelier** states that *when a change is made in an equilibrium system, net reaction will occur to compensate for the change.* Consider the generalized reaction in equation (2) as an example. Suppose that an extra quantity of product species C is added to the equilibrium system:

$$A + B \underset{\text{net reaction occurs}}{\overset{\rightarrow}{\longleftarrow}} \overset{\downarrow \text{excess C added}}{C} + D \qquad (4)$$

This change in concentration conditions will result in a net chemical reaction in the reverse direction. Species A and B will be produced at the expense of C and D until a new state of equilibrium is reached.

2.2 ACIDS AND BASES

Definitions of Acids and Bases

The concept of acids and bases arises from a need to describe and understand a wide variety of reactions. Acids and bases may be defined in several ways, depending on the types of reactions involved. Two approaches are useful for the interpretation of cation and anion reactions in water solution. The **Brønsted–Lowry concept** is used to describe proton-transfer reactions, and the **Lewis concept** is used to describe the formation of complex ions. According to the Brønsted–Lowry concept, an acid is a species that donates a proton and a base is a species that accepts a proton. The use of definitions that emphasize the role of the proton, H^+, are appropriate for water-solvent systems.

Several examples of Brønsted–Lowry acids and bases are shown in Table 2.1. Each of the species HCl, $HC_2H_3O_2$, and $Al(H_2O)_6^{3+}$ is an acid, because each species can donate a proton to a water molecule. The species NH_3 and CO_3^{2-} are bases, because they accept a proton from a water molecule. The common base, $NaOH$, dissociates in aqueous solution to produce OH^- directly. The Brønsted–Lowry definition

Table 2.1 | *Brønsted–Lowry Acids and Bases in Water*

Acids	Bases
$HCl + H_2O \rightarrow H_3O^+ + Cl^-$	$NH_3 + H_2O \rightleftarrows NH_4^+ + OH^-$
$HC_2H_3O_2 + H_2O \rightleftarrows H_3O^+ + C_2H_3O_2^-$	$CO_3^{2-} + H_2O \rightleftarrows HCO_3^- + OH^-$
$Al(H_2O)_6^{3+} + H_2O \rightleftarrows H_3O^+ + Al(H_2O)_5OH^{2+}$	

enlarges the concept of a base. NH_3 and CO_3^{2-} are bases even though they do not contain hydroxide that is released directly into solution.

Some acids are able to donate more than one proton. Diprotic acids such as H_2SO_4 and H_2S can donate either one or two protons, depending on the strength and concentration of the bases present.

A Brønsted–Lowry acid–base reaction involves a proton transfer. Consider the reaction of $HC_2H_3O_2$ with H_2O:

$$HC_2H_3O_2 + H_2O \rightleftarrows H_3O^+ + C_2H_3O_2^- \tag{5}$$

A proton is transferred from $HC_2H_3O_2$ to H_2O. In the process a new acid, H_3O^+, and a new base, $C_2H_3O_2^-$, are produced. In the reverse reaction in equation (5), H_3O^+ transfers a proton back to $C_2H_3O_2^-$. The H_3O^+ ion is referred to as the conjugate acid of H_2O; the $C_2H_3O_2^-$ ion is the conjugate base of $HC_2H_3O_2$. The proton in aqueous solution is present as the hydronium ion, H_3O^+. This ion, in turn, probably is hydrated by three molecules of water, $H_3O(H_2O)_3^+$. For simplicity the hydronium ion often is represented simply as H^+.

The Lewis acid–base concept is also useful for the interpretation of cation and anion reactions. This concept focuses attention on the formation of an electron-pair bond. A Lewis acid is an electron-pair acceptor; a Lewis base is an electron-pair donor. Equation (6), involving an acid, A, and a base, B, illustrates these ideas:

$$A + :B \rightarrow A:B \tag{6}$$

The base species has one or more unshared pairs of electrons. The acid species must possess one or more empty orbitals, capable of bond formation. The reaction of a proton (having an empty $1s$ orbital) with an ammonia molecule (having an unshared pair of electrons) is an example of a Lewis acid–base reaction:

$$H^+ + :NH_3 \rightarrow NH_4^+ \tag{7}$$

All Brønsted–Lowry bases are Lewis bases also. Base species such as NH_3, OH^-, and CO_3^{2-} all have unshared pairs of electrons which form a strong bond with a proton. The Lewis definition enlarges the concept of an acid. Not only protons, but also all other species with unfilled orbitals, are potentially acids. The reaction of a cation to form a complex ion is an important example of a Lewis acid–base reaction (see Section 2.5).

Dissociation of Acids and Bases

The process of transfer of a proton from an acid to a solvent water molecule is a dissociation reaction. Various acids differ in their ability to donate a proton. Acids that readily donate a proton in aqueous solution are termed **strong acids**. (The adjective strong refers to donating ability, not to concentration.) Examples of common strong acids are HCl, HNO_3, $HClO_4$, and H_2SO_4 (first proton). These acids are completely dissociated into ions in dilute aqueous solution. Thus, a $0.1\ M$ HCl solution contains $0.1\ M\ H^+$ and $0.1\ M\ Cl^-$, the concentration of HCl molecules being negligibly small. Acids that have only a small tendency to donate a proton are referred to as **weak acids**. Examples of common weak acids are HF, $HC_2H_3O_2$, $HOCl$, and CO_2 (or H_2CO_3). The partial dissociation of $HC_2H_3O_2$ is described by equation (3). A $0.1\ M\ HC_2H_3O_2$ solution is only about 1 percent dissociated into ions. Thus, $HC_2H_3O_2$ molecules are the predominant species in such a solution.

Bases also vary in their ability to accept a proton. Strong bases, such as $NaOH$ and $Ba(OH)_2$, are completely dissociated into hydroxide ions and the corresponding cation. The resulting OH^- ions are very strong proton acceptors. Ammonia is a weak base. It produces

$$NH_3 + H_2O \rightleftarrows NH_4^+ + OH^- \tag{8}$$

only a low concentration of hydroxide ion in solution. The extent of dissociation of acids and bases is treated quantitatively in Chapter 4.

Cations React as Acids

An important characteristic of a number of cations is their ability to react as acids in water solution. This characteristic is most pronounced for ions such as Al^{3+} that are small and have a high positive charge. Solutions of most aluminum salts contain hydrogen ion in excess of that expected for a neutral solution. The aluminum ions are present in water as the hydrated species $Al(H_2O)_6^{3+}$. Because of the close proximity of the $+3$ charge on aluminum, a proton can be transferred readily from one of the bound H_2O molecules:

$$Al(H_2O)_6^{3+} + H_2O \rightleftarrows H_3O^+ + Al(H_2O)_5OH^{2+} \tag{9}$$

Although an appreciable concentration of hydrogen ion is produced, aluminum(III) is probably best classified as a weak acid. Other $+3$ cations undergo similar reactions. However, such a reaction is far less important for the $+1$ and $+2$ ions. Reaction (9) and the corresponding reaction of anions as bases are sometimes called **hydrolysis** (water-splitting) **reactions**. It seems more appropriate, though, to focus attention on dissociation to yield a proton, rather than on the splitting of a water molecule.

Anions React as Bases

Any anion can, in principle, be a Brønsted–Lowry base. Anions are hydrated in water solution. The stronger base anions accept a proton from a neighboring water molecule, releasing an OH^- ion to the solution. For example, solutions of $NaC_2H_3O_2$ contain an excess of OH^- because of the reaction

$$C_2H_3O_2^- + H_2O \rightleftarrows HC_2H_3O_2 + OH^- \qquad (10)$$

Note that this reaction is very much like that in equation (8), the base dissociation of NH_3. Sulfide ion is a considerably stronger base than either $C_2H_3O_2^-$ or NH_3:

$$S^{2-} + H_2O \rightleftarrows HS^- + OH^- \qquad (11)$$

Solutions of Na_2S are exceedingly basic because of the existence of reaction (11). Thus, one useful means of identifying S^{2-} in solution is to test the basicity of the solution. Some common strong-base anions are S^{2-}, CO_3^{2-}, and PO_4^{3-}. Anions such as $C_2H_3O_2^-$, F^-, and NO_2^- are weak bases. Species such as NO_3^-, ClO_4^-, Cl^-, and Br^- have negligible base character.

A question naturally arises as to how one can tell whether or not a given anion will exhibit appreciable basicity. A preliminary answer can be made by considering the strength of the conjugate acid. For example, HNO_3 and HCl are very strong acids. The corresponding bases, NO_3^- and Cl^-, are very weak. An inverse relationship exists. Since the conjugate acids of S^{2-} and CO_3^{2-} are very weak, these anions are quite basic.

If attention is restricted to oxyanions, it becomes possible to understand base strength in a rather general way. For example, compare carbonate and nitrate anions. Carbonate ion is a reasonably strong base; nitrate is a very weak base. Both CO_3^{2-} and NO_3^- are composed of a central atom bonded to three oxygen atoms. A significant difference is that the -2 charge on carbonate is twice as great as the -1 charge on nitrate. The quantity **negative charge per oxygen atom** for CO_3^{2-} is $-\frac{2}{3} = -0.67$; similarly, the negative charge per oxygen atom for NO_3^- is $-\frac{1}{3} = -0.33$:

-0.67 charge unit per oxygen atom
strong base

-0.33 charge unit per oxygen atom
very weak base

When transferred to an anion, a proton becomes bonded to a partially negative oxygen atom. Therefore, the greater the quantity negative charge per oxygen atom, the greater the basicity of the anion. These ideas can be extended to other oxyanions also. A series of oxyanions is compared in Table 2.2. In this table the anions are grouped together according to values of the ratio negative charge per oxygen atom. The conjugate acid of each anion is also shown, along with the corresponding values of the acid-dissociation constants. As these K_a values become smaller, the base strength of the anion increases. From Table 2.2 it is

Table 2.2 | *Base Strength of Anions*

Negative charge per oxygen atom	Anion		Conjugate acid	K_a for the conjugate acid
-1.00	ClO^-	↑	$HClO$	10^{-8}
-0.67	$\begin{cases} CO_3^{2-} \\ SO_3^{2-} \end{cases}$	base	$\begin{matrix} HCO_3^- \\ HSO_3^- \end{matrix}$	$\begin{matrix} 10^{-11} \\ 10^{-7} \end{matrix}$
-0.50	$\begin{cases} C_2H_3O_2^- \\ NO_2^- \\ ClO_2^- \\ H_2PO_2^- \end{cases}$	strength increases	$\begin{matrix} HC_2H_3O_2 \\ HNO_2 \\ HClO_2 \\ H_3PO_2 \end{matrix}$	$\begin{matrix} 10^{-5} \\ 10^{-4} \\ 10^{-2} \\ 10^{-2} \end{matrix}$
-0.33	$\begin{cases} ClO_3^- \\ NO_3^- \\ HSO_4^- \end{cases}$		$\begin{matrix} HClO_3 \\ HNO_3 \\ H_2SO_4 \end{matrix}$	$\begin{matrix} >1 \\ >1 \\ >1 \end{matrix}$
-0.25	ClO_4^-		$HClO_4$	$\gg 1$

apparent that the base strength of anions increases as the corresponding value of negative charge per oxygen atom increases. This general rule is a useful one to remember. However, the correlation is by no means exact, because several factors (such as changes in the solvent) have been ignored.

2.3 | SALTS OF LOW SOLUBILITY. PRECIPITATION REACTIONS

Solubility in Water

Solutes that dissolve in water to form solutions of appreciable concentration are referred to as being soluble in water. Each solute is characterized by its **solubility**, which is defined to be the maximum concentration of solution that can be produced under a given set of conditions. A solution is **saturated** when it contains the maximum concentration of solute. Salts vary widely in solubility. For example, the solubility of NaCl in water is approximately 3.5 M, compared with a value of only 10^{-5} M for AgCl. The presence of excess solid solute in contact with a saturated solution constitutes an equilibrium system. When solid AgCl is added to a quantity of water, AgCl dissolves until the resulting solution is saturated at 10^{-5} mole of AgCl per liter of solution. Once saturation is reached, there is no further net change in the concentration of the solution or of the quantity of solid remaining. Dynamic equilibrium exists, the rate at which AgCl dissolves being exactly equal to the rate at which Ag^+ and Cl^- precipitate:

$$AgCl(s) \rightleftarrows Ag^+ + Cl^- \qquad (12)$$

Like most salts, AgCl dissociates completely into ions as it dissolves.

Cations React with Anions to Form Salts of Low Solubility

Precipitation reactions are characteristic of many ions and are among the most important types of reactions that occur in water solution. A precipitation reaction is just the reverse of a solubility equilibrium reaction. The silver chloride system can be used to illustrate this point. When solutions of $AgNO_3$ and HCl are mixed, AgCl precipitates until the concentrations of Ag^+ and Cl^- decrease to their equilibrium values:

$$Ag^+ + Cl^- \rightarrow AgCl(s)* \tag{13}$$

(This point is discussed quantitatively in Section 4.3.) An $AgNO_3$ solution contains Ag^+ and NO_3^- ions, stabilized by solvent water molecules. Similarly, H^+ and Cl^- ions are strongly hydrated by water molecules in an HCl solution. When mixed, the Ag^+ and Cl^- ions interact with each other, forming a solid with very strong bonds between the ions. In the case of silver chloride, precipitation occurs because the lattice energy of the solid exceeds the hydration energy of the ions in solution.

Effect of Acidity on the Solubility of Salts

The solubility of some salts is dependent on the hydrogen-ion concentration in solution. This occurs when the anion is appreciably basic. As an example, consider the solubility of barium sulfite in aqueous solution:

$$Ba^{2+} + SO_3^{2-} \rightleftarrows BaSO_3(s) \tag{14}$$

$$H^+ + SO_3^{2-} \rightleftarrows HSO_3^-(aq) \tag{15}$$

In neutral or basic solution, $BaSO_3$ precipitates when Na_2SO_3 and $BaCl_2$ are mixed. However, $BaSO_3$ is not formed in a highly acidic solution. Also, addition of acid to a $BaSO_3$ precipitate causes the precipitate to dissolve:

$$BaSO_3(s) + H^+ \rightarrow Ba^{2+} + HSO_3^- \tag{16}$$

These observations can be understood by recognizing that H^+ and Ba^{2+} must compete for SO_3^{2-} ions. In acid solution, hydrogen ion is present in large enough concentration to tie up most of the sulfite ions as HSO_3^-. As a result, $BaSO_3$ does not form.

The solubility of barium sulfate, $BaSO_4$, is not affected by the acidity of the solution. Sulfate ion is not very basic; thus, barium ions compete much more strongly than do the hydrogen ions for sulfate.

2.4 | OXIDATION–REDUCTION REACTIONS

Oxidation and Reduction

An oxidation–reduction reaction is one of the important types of reactions that occur commonly in water solution. Oxidation and reduction are not separate processes; they always occur together. *Oxidation can be defined as a process in which an element undergoes an increase in oxidation number. Reduction is a process in which an element*

---

* (s), solid; (l), liquid; (g), gas; (aq), aqueous.

undergoes a decrease in oxidation number. Oxidation numbers are assigned according to the following rules.

1. Each atom of an element is assigned an oxidation number of zero:

    Examples: $\overset{(0)}{\text{Na}}\quad \overset{(0)}{\text{N}_2}\quad \overset{(0)}{\text{P}_4}$

2. The oxidation number of a monatomic ion is taken as the charge on that ion:

    Examples: $\overset{(-1)}{\text{Cl}^-}\quad \overset{(+2)}{\text{Fe}^{2+}}\quad \overset{(+3)}{\text{Al}^{3+}}$

3. The oxidation number of hydrogen is $+1$ in hydrogen-containing compounds. (However, it is -1 in metallic hydrides, such as NaH.) The oxidation number of oxygen is -2 in oxygen-containing compounds. (However, it is -1 in peroxides, such as H_2O_2.)

    Examples: $\overset{(+1)}{\text{H}^+}\quad \overset{(+1)(-2)}{\text{H}_2\text{O}}\quad \overset{(+1)}{\text{CH}_4}\quad \overset{(-2)}{\text{O}^{2-}}$

4. For a neutral molecule, the algebraic sum of oxidation numbers for all the atoms must equal zero. For an ion, the algebraic sum of oxidation numbers for all the atoms must equal the charge on the ion.

    Examples:

    $\overset{(+1)(-2)}{\text{H}_2\text{O}}\qquad 2(1) + (-2) = 0$

    $\overset{(+1)(+6)(-2)}{\text{H}_2\text{SO}_4}\qquad 2(1) + (6) + 4(-2) = 0$

    $\overset{(+4)(-2)}{\text{CO}_3^{2-}}\qquad (4) + 3(-2) = -2$

Note that in the examples above, the oxidation number of S is assigned as $+6$ and that of C as $+4$, based on a combination of rules 3 and 4. In the case of H_2SO_4, the number for S is obtained as the solution of the equation $2(1) + x + 4(-2) = 0$, in which $x = 6$.

Equation (17) is an example of a common oxidation–reduction reaction:

$$\overset{(+4)}{\text{H}\underline{\text{S}}\text{O}_3^-} + \overset{(0)}{\underline{\text{I}}_2} + \text{H}_2\text{O} \rightarrow \overset{(+6)}{\text{H}\underline{\text{S}}\text{O}_4^-} + \overset{(-1)}{2\underline{\text{I}}^-} + 2\text{H}^+ \qquad (17)$$

The oxidation numbers of the underlined atoms change in the reaction. The HSO_3^- is oxidized by I_2, which is referred to as the **oxidizing agent**. At the same time, I_2 is reduced by HSO_3^-, which is referred to as the **reducing agent**.

Cations and Anions React as Oxidizing or Reducing Agents

Many of the common oxidizing or reducing agents in water solution are cations or anions. The characteristic reactions which these ions undergo often are oxidation–reduction reactions. A major objective of the cation and anion experiments in Parts II and III is the examination of some of these reactions.

The oxidation–reduction behavior of an ion depends on (1) its relative reduction potential (discussed below), (2) the oxidation state of the principal element present, (3) conditions in solution, and (4) the rate at which a given oxidation–reduction reaction can occur. As examples, consider the elements iron and sulfur:

$$\underset{(+3)}{Fe^{3+}} \quad \underset{(+2)}{Fe^{2+}} \quad \underset{(0)}{Fe}$$

$$\underset{(+6)}{\underline{S}O_4^{2-}} \quad \underset{(+4)}{\underline{S}O_3^{2-}} \quad \underset{(0)}{S} \quad \underset{(-2)}{S^{2-}}$$

In the sulfide anion, S^{2-}, sulfur is in its lowest possible oxidation state (-2). (An element is often referred to as being in an oxidation state, numerically equal to the oxidation number.) Species in the lowest oxidation state are frequently, although not necessarily, reducing agents. Sulfide ion is a typical example of a reducing agent. The ability to reduce a wide variety of oxidizing agents is characteristic of S^{2-}. The $+3$ state is the highest normal oxidation state of iron. Iron(III) is a typical example of an oxidizing cation. When Fe^{3+} and S^{2-} are mixed in acid solution (in which sulfide is present as H_2S), the following reaction takes place:

$$2Fe^{3+} + H_2S \rightarrow 2Fe^{2+} + S + 2H^+ \qquad (18)$$

This reaction is important in the Group 2 and Group 3 chemistry of cations in Part II.

From the above list of oxidation states of sulfur, it might be expected that SO_4^{2-} ion should be an oxidizing agent. Sulfate ion can function as a weak oxidizing agent but only in concentrated acid at higher temperatures. In judging whether or not certain reactions will be observed, both the tendencies of the reactions to occur (thermodynamic spontaneity) and the rates of the reactions must be considered. Thus, it is important to remember that not all oxidizing (or reducing) agents react rapidly. Sulfate ion reacts rapidly only in highly acid solution. The perchlorate ion, ClO_4^-, provides a striking example of this point. Although ClO_4^- potentially is a very strong oxidizing agent, it will not oxidize other substances in dilute aqueous solution. Only in concentrated acid does ClO_4^- function as an oxidizing agent.

Spontaneous Oxidation–Reduction Reactions

An important area in the study of ions in aqueous solution is the prediction of which oxidation–reduction reactions can occur spontaneously. Such predictions can be made using a table of standard reduction potentials. Shown in Table 2.3 are a series of half-reactions that relate two different oxidation states of a given element. In the first half-reaction listed,

$$Na^+ + e^- \rightleftarrows Na \qquad E^0 = -2.71 \text{ volts}$$

sodium in the $+1$ state is reduced to metallic sodium in the 0 oxidation state by addition of an electron (e^-). This process cannot occur by itself but only when combined with another half-reaction. The numerical voltage quantity, E^0, is the standard reduction potential of sodium. The standard value of -2.71 volts refers to unit concentrations of all species at 25°C. All the standard potentials are arbitrary numbers based on the definition that $E^0 = 0.00$ volts for the hydrogen ion–hydrogen gas half-reaction. Each of the principal chemical species listed on the left side of Table 2.3 is a potential oxidizing agent. Correspondingly, each of the principal species listed on the right side is a potential reducing agent. The strongest oxidizing agents are found at the bottom (left) of the table: for example, MnO_4^- and BrO_3^-. The strongest reducing agents are found at the top (right) of the table: for example, Na and Zn.

The appearance of electrons in the half-reactions suggests that half-reactions may be combined to produce a cell capable of delivering an electric current. Such electrochemical cells especially constructed to produce current in an external circuit are often referred to as **galvanic cells**. These cells are discussed in detail in your general chemistry textbook, which should be consulted for further information. It is possible to measure the difference in potential, E^0_{cell}, between two electrodes at which the two half-reactions occur. At standard conditions,

Table 2.3 | *Standard Reduction Potentials*

Half-reaction	E^0 (volts)
$Na^+ + e^- \rightleftarrows Na$	-2.71
$Zn^{2+} + 2e^- \rightleftarrows Zn$	-0.76
$Fe^{2+} + 2e^- \rightleftarrows Fe$	-0.41
$Cd^{2+} + 2e^- \rightleftarrows Cd$	-0.40
$HSO_4^- + 3H^+ + 2e^- \rightleftarrows H_2SO_3 + H_2O$	-0.17
$Sn^{2+} + 2e^- \rightleftarrows Sn$	-0.14
$2H^+ + 2e^- \rightleftarrows H_2$	0.00
$S_4O_6^{2-} + 2e^- \rightleftarrows 2S_2O_3^{2-}$	0.10
$CrO_4^{2-} + 4H_2O + 3e^- \rightleftarrows Cr(OH)_4^- + 4OH^-$ (basic solution)	0.12
$S + 2H^+ + 2e^- \rightleftarrows H_2S$	0.14
$Sn^{4+} + 2e^- \rightleftarrows Sn^{2+}$	0.14
$Cu^{2+} + 2e^- \rightleftarrows Cu$	0.34
$I_2 + 2e^- \rightleftarrows 2I^-$ (or $I_3^- + 2e^- \rightleftarrows 3I^-$)	0.54
$O_2 + 2H^+ + 2e^- \rightleftarrows H_2O_2$	0.68
$Fe^{3+} + e^- \rightleftarrows Fe^{2+}$	0.77
$Hg_2^{2+} + 2e^- \rightleftarrows 2Hg$	0.80
$Ag^+ + e^- \rightleftarrows Ag$	0.80
$O_2^{2-} + 2H_2O + 2e^- \rightleftarrows 4OH^-$ (basic solution)	0.88
$2Hg^{2+} + 2e^- \rightleftarrows Hg_2^{2+}$	0.91
$HNO_2 + H^+ + e^- \rightleftarrows NO + H_2O$	1.00
$O_2 + 4H^+ + 4e^- \rightleftarrows 2H_2O$	1.23
$Cr_2O_7^{2-} + 14H^+ + 6e^- \rightleftarrows 2Cr^{3+} + 7H_2O$	1.33
$Cl_2 + 2e^- \rightleftarrows 2Cl^-$	1.36
$BrO_3^- + 6H^+ + 6e^- \rightleftarrows Br^- + 3H_2O$	1.44
$MnO_4^- + 8H^+ + 5e^- \rightleftarrows Mn^{2+} + 4H_2O$	1.49

NOTE: E^0 values refer to unit concentrations and pressure in acid solution at 25°C. The values may be converted into standard oxidation potentials by changing the sign of each E^0 and reversing each half-reaction.

the voltage of the cell is equal to the difference in the standard reduction potentials for the two half-cells involved, $E_{cell}^0 = E_1^0 - E_2^0$. Combining half-reactions 1 and 2 results in an oxidation–reduction reaction. The half-reactions that comprise reaction (18) can be used to illustrate these ideas.

Half-reaction 1:

$$Fe^{3+} + e^- \rightleftarrows Fe^{2+} \qquad E_1^0 = 0.77 \text{ volt}$$

Half-reaction 2:

$$S + 2H^+ + 2e^- \rightleftarrows H_2S \qquad E_2^0 = 0.14 \text{ volt}$$

Combining the two half-reaction potentials gives $E_{cell}^0 = 0.77 - (0.14) = 0.63$ volt. Combining the two half-reactions results in the following cell reaction:

$$\begin{array}{c} 2(Fe^{3+} + e^- \rightleftarrows Fe^{2+}) \\ H_2S \rightleftarrows S + 2H^+ + 2e^- \\ \hline 2Fe^{3+} + H_2S \rightarrow 2Fe^{2+} + S + 2H^+ \end{array}$$

Note that one of the half-reactions must be reversed before the half-reactions are combined.

One half-reaction consumes electrons; the other half-reaction produces electrons. A positive value of E_{cell}^0 means that the cell reaction proceeds spontaneously from left to right as written. Therefore, the problem of predicting which oxidation–reduction reactions are spontaneous becomes one of deciding which reactions have $E_{cell}^0 > 0$. The half-reaction with the smaller (algebraic) value of E^0 should be reversed and, consequently, its standard potential subtracted. Referring back to the Fe^{3+}–H_2S reaction, it is clear that reversing half-reaction 1 would result in $E_{cell}^0 = -0.63$ volt and the nonspontaneous reaction

$$2Fe^{2+} + S + 2H^+ \rightleftarrows 2Fe^{3+} + H_2S \qquad (19)$$

Thus, using the standard potentials in Table 2.3, one can show that reaction (18) should be spontaneous. Results in Group 2 cation chemistry confirm this prediction. To further illustrate the usefulness of Table 2.3, consider the question of whether or not I_2 is a sufficiently strong oxidizing agent to convert sulfite ion to sulfate in acid solution. The required half-reactions are

$$HSO_4^- + 3H^+ + 2e^- \rightleftarrows H_2SO_3 + H_2O \qquad E^0 = -0.17 \text{ volt}$$

$$I_2 + 2e^- \rightleftarrows 2I^- \qquad E^0 = 0.54 \text{ volt}$$

Reversing the first half-reaction (which has the lower E^0 value) leads to reaction (20) with a corresponding $E_{cell}^0 = 0.71$ volt:

$$I_2 + H_2SO_3 + H_2O \rightarrow 2I^- + HSO_4^- + 3H^+ \qquad (20)$$

The oxidation of H_2SO_3 by I_2 is spontaneous and should be observable in the laboratory, provided that the rate of reaction (20) is not too low.

The technique of combining half-reactions also provides a useful means of balancing oxidation–reduction equations. This technique is referred to as the ion–electron method for balancing equations and is discussed in Section 3.2.

2.5 FORMATION OF COMPLEX IONS

A complex ion is a chemical combination of a metal ion surrounded by two or more anions or molecules called **ligands**. Two examples of complex ions that will be encountered in Parts II and III are

$$Fe(H_2O)_5NCS^{2+} \quad \text{and} \quad Ag(S_2O_3)_2^{3-}$$

The nature of these, and other complex ions, is discussed in detail in Chapters 5 and 6. In the complex on the left, a Fe^{3+} ion is bonded to five H_2O molecules and one SCN^- ion. Two $S_2O_3^{2-}$ anions are the ligands that are bonded to the Ag^+ ion in the complex ion on the right. Reactions that lead to the formation of complex ions are important because such reactions characterize the behavior of many cations and anions. The complex ions often have distinctive colors which make them useful in identifying a particular cation or anion. A good example is the identification of iron(III) with thiocyanate ion. When a thiocyanate salt is added to a solution containing Fe^{3+}, the color changes from light yellow to a deep red:

$$Fe(H_2O)_6^{3+} + SCN^- \rightarrow Fe(H_2O)_5NCS^{2+} + H_2O \qquad (21)$$

The formation of the $Fe(H_2O)_5NCS^{2+}$ complex ion in reaction (21) is responsible for the red color. Complex ions are often soluble in dilute aqueous solution. Thus, failure to observe precipitation when a cation is added to an anion does not necessarily mean that no reaction occurred. Complex-ion formation may have taken place instead. The addition of Ag^+ to a thiosulfate solution provides an interesting example. If a low concentration of Ag^+ is added, complex-ion formation occurs:

$$Ag^+ + 2S_2O_3^{2-} \rightarrow Ag(S_2O_3)_2^{3-}(aq) \qquad (22)$$

Since the $Ag(S_2O_3)_2^{3-}$ complex ion is colorless, it is not obvious that any net change occurs. If the concentration of Ag^+ is high so that $S_2O_3^{2-}$ is no longer in excess, precipitation occurs instead:

$$2Ag^+ + S_2O_3^{2-} \rightarrow Ag_2S_2O_3(s) \qquad (23)$$

These and other reactions of complex ions will be explored in detail in Parts II and III.

Problems

1. In what direction (to the left, or to the right) will net reaction occur due to the change indicated in each equilibrium system?

Equilibrium Reaction	Change	Direction of Net Reaction
a. $H_2(g) + I_2(g) \rightleftarrows 2HI(g)$	I_2 added	_____
b. $CaCO_3(s) \rightleftarrows CaO(s) + CO_2(g)$	CO_2 removed	_____
c. $HC_2H_3O_2 + H_2O \rightleftarrows H_3O^+ + C_2H_3O_2^-$	$C_2H_3O_2^-$ added	_____
d. $AgCl(s) \rightleftarrows Ag^+ + Cl^-$	Cl^- added	_____

2. Referring to *1c*, suggest a solute that could be added to an aqueous solution of $HC_2H_3O_2$ in order to decrease the concentration of hydronium ion.

3. Circle the species that are
   a. Brønsted–Lowry acids: H_2O, Mg^{2+}, SO_3^{2-}, HF
   b. Brønsted–Lowry bases: H_2O, $C_2H_3O_2^-$, NH_4^+, $Al(H_2O)_6^{3+}$

4. Write the formula of the conjugate base.

   HNO_2_____ NH_4^+_____ HCO_3^-_____

5. Write the formula of the conjugate acid.

   H_2O_____ HSO_4^-_____ CO_3^{2-}_____

6. Write a chemical equation that illustrates the reaction of
   a. ClO^- as a base.

   b. BF_3 as an acid. Show the Lewis electron-dot structures of each of the species.

7. Underline the stronger acid in each pair.

   $HClO_3$ $HClO_2$ HPO_4^{2-} HSO_4^-

   HSO_3^- HSO_4^- $HBrO_3$ HNO_2

8. Write the formulas of three silver salts which are insoluble in water but soluble in acid.

9. Determine the oxidation number of the underlined atom.

   $\underline{N}H_3$ $\underline{P}_4O_{10}$ $K_2\underline{S}_3O_6$ $\underline{C}_2H_5OH$ $H_2\underline{P}O_2^-$

10. In each reaction, indicate which species is the oxidizing agent and which the reducing agent.

    $$Fe^{2+} + HNO_2 + H^+ \rightarrow Fe^{3+} + NO + H_2O$$

    $$2MnO_4^- + 3HCOOH + 2H^+ \rightarrow 2MnO_2 + 3CO_2 + 4H_2O$$

$$3I_2 + 6OH^- \rightarrow 5I^- + IO_3^- + 3H_2O$$

11. Write the formula of two species capable of
  a. Oxidizing Fe^{2+} to Fe^{3+}.

  b. Reducing HSO_4^- to H_2SO_3.

12. Calculate E^0_{cell} for each reaction indicated and state whether or not the reaction will be spontaneous.
  a. $Ag^+ + Sn^{2+} \rightarrow$

  b. $HNO_2 + Cl^- \rightarrow$

  c. $Zn + Cd^{2+} \rightarrow$

Writing Chemical Equations

CHAPTER 3

The major types of reactions of cations and anions discussed in Chapter 2 form the basis of the experimental work in Parts II and III. The purpose of a chemical equation is to describe, in a very concise way, the nature of a particular chemical reaction. A useful equation should show the actual chemical change that takes place and the correct formulas of the products and reactants involved. In addition, the equation indicates the stoichiometry of the reaction by showing the relationship between the numbers of moles of reactants and products. The purpose of this chapter is to develop a set of general principles that will be useful in writing correct chemical equations. Most important is an understanding of the factors which make a spontaneous reaction possible and which determine the kinds of products that occur. The techniques of balancing chemical equations are reviewed and applied to a number of specific examples.

3.1 WRITING CORRECT PRODUCTS AND REACTANTS

Spontaneous Reactions

Chemical systems proceed spontaneously to states of lower energy and greater molecular randomness. Chemical thermodynamics involves a study of the energy changes in chemical reactions. Although this subject is beyond the scope of this book, it should be possible to recognize the types of products that occur in spontaneous reactions in solution. In general, such products are those formed in the major types of reactions of ions in solution, described in Chapter 2.

Nondissociated products. Many reactions proceed spontaneously to form a nondissociated product. The product may be a molecule of the solvent or some other molecular species. Two common examples are the formation of solvent H_2O in an acid–base reaction and the formation of a weak acid from a salt of the weak acid.

Insoluble substances. Reactions that lead to the formation of an insoluble salt or hydroxide are spontaneous. For example, a soluble silver salt, when mixed with a source of chloride ion, reacts to form insoluble silver chloride. Such reactions frequently go to completion (that is, all the reactants are converted to products) because the solid separates from the solution. Also, reactions in which a salt of low solubility is converted to another salt of even lower solubility are spontaneous.

Oxidation–reduction reactions. Reactions in which a stronger oxidizing agent produces a weaker oxidizing agent are spontaneous. (In every oxidation–reduction reaction, a new oxidizing agent, and a new reducing agent, are produced.) Spontaneous reactions are associated with positive values of E^0_{cell}, as discussed in Chapter 2.

Formation of a gaseous product. Reactions which lead to formation of a gas that escapes from solution are spontaneous. Such reactions often go to completion because of the loss of the gaseous product from solution. Common examples include the release of carbon dioxide from a carbonate salt, or of sulfur dioxide from a sulfite salt, in acid solution.

Complex-ion reactions. A spontaneous reaction can occur when one or more ligands in a complex ion are replaced by ligands that are more tightly bound. Addition of concentrated ammonia to a solution containing copper(II) leads to spontaneous replacement of water molecules around the Cu^{2+} ion by ammonia molecules.

Writing Correct Products and Reactants

The rules listed below should serve as a useful guide in predicting reaction products. *Making sensible choices of products, under the conditions specified, is the most important part of writing chemical equations.*

1. ***Correct formulas.*** Molecular formulas should conform with bonding rules. All ionic species should be written with the proper charge (see Table 1.1).
2. ***Reaction types.*** Decide whether the reaction involves oxidation–reduction or a simple interchange of ions. Oxidation–reduction is probable if one of the reactants is a strong oxidizing agent, a strong reducing agent, or is in the elemental state.
3. ***Products of spontaneous reactions.*** Select a product that is of the type usually found in spontaneous reactions.

Nondissociated species:	H_2O, $HC_2H_3O_2$, NH_4^+
Insoluble species:	$AgCl$, $BaSO_4$, $Al(OH)_3$
Complex ion:	$Fe(H_2O)_5NCS^{2+}$, $Ag(S_2O_3)_2^{3-}$
Gaseous product:	CO_2, SO_2, H_2S

   (Only rarely are H_2 and O_2 products of a reaction in aqueous solution. *Do not write H_2 or O_2 as products simply to balance hydrogen or oxygen in an equation.*)

   Products in different oxidation states: Fe^{3+} reduced to Fe^{2+}, HSO_3^- oxidized to HSO_4^-, HNO_3 reduced to NO.

   (In the last example, nitrogen in HNO_3 might also have been reduced to several other oxidation states, ranging from $+4$ to -3. In a case such as this, experimental tests or other outside evidence is required to decide which oxidation state to choose. If such information is unavailable, make a tentative choice of a reasonable state.)
4. ***Reaction conditions.*** The products must be compatible with the reaction conditions. Basic species cannot be products in an acid

solution; for example, write NH_4^+ rather than NH_3, or H_2O rather than OH^-, in acid solution. Acidic species cannot be products in basic solution; for example, write SO_4^{2-} rather than H_2SO_4. Reducing species, such as Zn, Fe^{2+}, and H_2S, cannot be produced in a solution containing a strongly oxidizing reactant such as $Cr_2O_7^{2-}$, O_2, or Cl_2. The converse is also true; oxidizing species will not be produced in a solution containing a strong reducing agent.

5. **Complete reactions.** The product selected should be stable and not subject to further reaction under the conditions. For example, H_2SO_4 is neutralized to SO_4^{2-}, not HSO_4^-, when excess base is added.

3.2 BALANCING CHEMICAL EQUATIONS

Reactions of Ions and Molecules Not Involving Oxidation–Reduction

Reactions of this type do not involve changes in oxidation states of any of the reactants. Equations for these reactions can be written using the following steps:

1. Write formulas for the products.
2. Balance atoms by adjusting coefficients in front of reactants and products. (Do not change subscripts within a given formula.)
3. Balance charges so that the same *net* charge appears on each side of the equation.

Equations for reactions in water solution are usually written in **net-ionic** form. The purpose of the net-ionic equation is to show the product and reactant species present in solution and to show the actual chemical change which takes place. In water solution, substances that are primarily dissociated into ions should be written in the form of the separate ions. Ions that do not undergo chemical change are canceled on both sides of the equation. Net-ionic equations are simpler and more concise than molecular equations. Consider the reaction of acetic acid with sodium hydroxide. The products are written by interchanging the roles of the Na^+ and H^+ ions.

Molecular equation:

$$HC_2H_3O_2 + NaOH \rightarrow H_2O + NaC_2H_3O_2 \qquad (1)$$

Net-ionic equation:

$$HC_2H_3O_2 + Na^+ + OH^- \rightarrow H_2O + Na^+ + C_2H_3O_2^-$$
$$HC_2H_3O_2 + OH^- \rightarrow H_2O + C_2H_3O_2^- \qquad (2)$$

Because NaOH is 100 percent dissociated in water, it is written in the form of ions. However, $HC_2H_3O_2$ is not appreciably dissociated and, therefore, is written in molecular form. In general, acids with K_a values of 10^{-2} or smaller should be written in molecular form. (See Table 2.2.) The Na^+ ions are canceled because they do not undergo any chemical change. It is $HC_2H_3O_2$ and OH^- that undergo chemical change and are consumed.

Having written equation (2), it is important to make certain that it represents a spontaneous reaction, based on the criteria in Section 3.1. Note that in equation (2) the weaker base, $C_2H_3O_2^-$, has been produced by the stronger base, OH^-. Also, the product H_2O is less dissociated than is the reactant $HC_2H_3O_2$. These two qualitative factors strongly suggest that reaction (2) should be spontaneous.

Examples
HCl is neutralized with NaOH:

$$HCl + NaOH \rightarrow H_2O + NaCl \quad \text{(molecular)}$$
$$H^+ + OH^- \rightarrow H_2O \quad \text{(net-ionic)}$$

The species HCl, NaOH, and NaCl are all completely dissociated and, therefore, written in ionic form. The reaction is spontaneous because of the formation of nondissociated H_2O molecules. Na^+ and Cl^- do not undergo chemical change. The net-ionic equation for the reaction of any strong acid with any strong base is just the combination of H^+ and OH^- indicated above.

HCl is added to solid $BaCO_3$:

$$2HCl + BaCO_3(s) \rightarrow H_2O + CO_2 + BaCl_2 \quad \text{(molecular)}$$
$$2H^+ + BaCO_3(s) \rightarrow H_2O + CO_2 + Ba^{2+} \quad \text{(net-ionic)}$$

Since $BaCO_3$ has a low solubility in water, it is written in the form of the solid. The protons remove CO_3^{2-} from Ba^{2+} to produce nondissociated H_2O and CO_2 gas.

$AgNO_3$ reacts with H_2S:

$$2AgNO_3 + H_2S \rightarrow Ag_2S(s) + 2HNO_3 \quad \text{(molecular)}$$
$$2Ag^+ + H_2S \rightarrow Ag_2S(s) + 2H^+ \quad \text{(net-ionic)}$$

H_2S is a very weak acid and is written in molecular form. The reaction is spontaneous because of the formation of highly insoluble Ag_2S.

Reactions Involving Oxidation–Reduction

By definition, oxidation–reduction reactions involve changes in oxidation numbers. If the identity of one or more of the products is known, begin by assigning oxidation numbers to ensure that the reaction does involve changes in oxidation numbers. If the products are at first unknown, use the ideas in Section 3.1 to select a reasonable set of products. Remember that in each reaction one element must undergo an increase in oxidation number and one element must undergo a decrease in oxidation number. To complete and balance the equation, either of the systematic methods below can be used. Writing the equation in net-ionic form is preferred.

Method of Changes in Oxidation Numbers

1. Write out the equation showing only the principal reactants and products. Assign oxidation numbers to the key elements (usually not hydrogen or oxygen).
2. Indicate the number of electrons (e^-) lost per atom of the element which undergoes an increase in oxidation number. Similarly, indicate the number of electrons gained per atom of the element which undergoes a decrease in oxidation number. For example, if an atom of an element undergoes a change from the $+5$ state to the $+2$ state, it is assumed that a gain of $3e^-$ will bring about this change. (Actual electron transfer may or may not occur in a particular oxidation–reduction reaction. Electron transfer is used here as part of the technique for balancing equations.)
3. Indicate the number of electrons lost or gained per formula unit. This step is required only when there are two or more atoms of the same element in a formula unit: for example, two Cr atoms in $Cr_2O_7^{2-}$.
4. Make the number of electrons lost equal the number of electrons gained by placing appropriate integers before each of the reactants. (Subscripts within a formula unit are not changed.) If reactant A loses n_A electrons and reactant B gains n_B electrons,

$$A + B \longrightarrow$$
$$\text{loss of } n_A e^- \downarrow \quad \uparrow \text{gain of } n_B e^-$$

place the integer n_A in front of B and the integer n_B in front of A. If n_A and n_B have a common divisor, divide both n_A and n_B by this number.

5. Balance the number of atoms of each element in the equation, beginning with the elements present in the principal products and reactants. If the reaction occurs in *acid solution, add H^+ and/or H_2O* to either side of the equation, as required. In *basic solution, add OH^- and/or H_2O*, as required. Complete balancing of the atoms.
6. Balance charges so that the same *net* charge appears on each side of the equation. If steps 1 through 5 have been carried out properly, the charges should already balance. This last step can therefore be used as a check on the final equation. The order of steps 5 and 6 may be reversed, if desired. This is recommended for reactions that occur in basic solution.

Example. Balance the equation for the conversion of Fe^{2+} to Fe^{3+} by $Cr_2O_7^{2-}$ in acid solution.

1. A skeleton equation is written which includes the reactants and products given:

$$\underset{(+6)}{\underline{Cr}_2O_7^{2-}} + \underset{(+2)}{Fe^{2+}} \rightarrow \underset{(+3)}{Fe^{3+}} + \underline{?}$$

From the oxidation numbers assigned, Fe^{2+} is being oxidized; therefore, Cr in $Cr_2O_7^{2-}$ must be reduced to an oxidation state lower than $+6$. Because chromium(III) is a very stable and common oxidation state for chromium, Cr^{3+} is a reasonable choice for the remaining product:

$$\underset{(+6)}{\underline{Cr_2O_7}^{2-}} + \underset{(+2)}{Fe^{2+}} \to \underset{(+3)}{Fe^{3+}} + \underset{(+3)}{Cr^{3+}}$$

2. Each Fe^{2+} loses $1e^-$ and each Cr gains $3e^-$:

$$\underset{(+6)}{Cr_2O_7^{2-}} + \underset{(+2)}{Fe^{2+}} \to \underset{(+3)}{Fe^{3+}} + \underset{(+3)}{Cr^{3+}}$$

gain of $3e^-$ per Cr ↑ ↓ loss of $1e^-$

3. Each $Cr_2O_7^{2-}$ gains $6e^-$ per $Cr_2O_7^{2-}$ formula unit:

$$\underset{(+6)}{Cr_2O_7^{2-}} + \underset{(+2)}{Fe^{2+}} \to \underset{(+3)}{Fe^{3+}} + \underset{(+3)}{Cr^{3+}}$$

gain of $6e^-$ per $Cr_2O_7^{2-}$ ↑ ↓ loss of $1e^-$

4. The electron loss by Fe^{2+} ions can be made equal to the gain of $6e^-$ for $Cr_2O_7^{2-}$ by placing a 6 before Fe^{2+}:

$$Cr_2O_7^{2-} + 6Fe^{2+} \to Fe^{3+} + Cr^{3+}$$

($6e^-$ lost = $6e^-$ gained)

5. $$Cr_2O_7^{2-} + 6Fe^{2+} \to 6Fe^{3+} + 2Cr^{3+}$$

To complete the balancing, it is necessary to add H_2O to the right side so that oxygen appears on both sides of the equation. In addition, H^+ must be added to the left side so that hydrogen appears on both sides of the equation:

$$Cr_2O_7^{2-} + 6Fe^{2+} + H^+ \to 6Fe^{3+} + 2Cr^{3+} + H_2O$$

Balancing hydrogen and oxygen results in the final equation:

$$\underline{Cr_2O_7^{2-} + 6Fe^{2+} + 14H^+ \to 6Fe^{3+} + 2Cr^{3+} + 7H_2O}$$

6. To check the steps above, note that the net charge on the left side, $1(-2) + 6(+2) + 14(+1) = \underline{+24}$, is equal to the net charge on the right side, $6(+3) + 2(+3) = \underline{+24}$, as required.

Example. Complete and balance the equation:

$$Cr(OH)_4^- + HO_2^- \to CrO_4^{2-} + H_2O \quad \text{(basic solution)}$$

1. $$\underset{(+3)}{\underline{Cr(OH)_4}^-} + \underset{(-1)}{\underline{HO_2}^-} \to \underset{(+6)}{\underline{CrO_4}^{2-}} + \underset{(-2)}{\underline{H_2O}}$$

2. $$\underset{(+3)}{Cr(OH)_4^-} + \underset{(-1)}{HO_2^-} \to \underset{(+6)}{CrO_4^{2-}} + \underset{(-2)}{H_2O}$$
loss of $3e^-$ ↓ ↑ gain of $1e^-$ per O atom

3. $$\underset{(+3)}{Cr(OH)_4^-} + \underset{(-1)}{HO_2^-} \to \underset{(+6)}{CrO_4^{2-}} + \underset{(-2)}{H_2O}$$
loss of $3e^-$ ↓ ↑ gain of $2e^-$ per HO_2^-

4. $$2Cr(OH)_4^- + 3HO_2^- \to CrO_4^{2-} + H_2O$$

6. In basic solution, apply steps 5 and 6 in reverse order:

$$2Cr(OH)_4^- + 3HO_2^- \rightarrow 2CrO_4^{2-} + H_2O$$

To balance charge, one OH^- must be added to the right side:

$$2Cr(OH)_4^- + 3HO_2^- \rightarrow 2CrO_4^{2-} + OH^- + H_2O$$

5. To balance hydrogen and oxygen, a total of five H_2O on the right side is required:

$$\underline{2Cr(OH)_4^- + 3HO_2^- \rightarrow 2CrO_4^{2-} + OH^- + 5H_2O}$$

Method of Half-Reactions

1. Write out the equation showing the principal products and reactants. Divide the equation into two half-reactions (Section 2.4). One half-reaction should involve reduction of an element from a higher to a lower oxidation state; the other should involve oxidation of an element.
2. Balance the number of each of the atoms in the two half-reactions, beginning with the principal element. If the reaction occurs in *acid solution*, *add H^+ and/or H_2O* to either side of the half-reaction, as required. In *basic solution, add OH^- and/or H_2O*, as required. Complete balancing of the atoms.
3. Balance charges in each half-reaction by adding electrons to the appropriate side of the half-reaction. The same *net* charge should appear on each side of the half-reaction. (Although oxidation numbers may be used to check the assignment of electrons, it is not necessary to use oxidation numbers.)
4. Multiply the coefficients in each of the half-reactions by an integer that will make the number of electrons lost in the oxidation process equal to the number of electrons gained in the reduction process. Add the two half-reactions and cancel any species common to both sides of the final equation. (The electrons must cancel and, therefore, do not appear in the final equation.)

Example. Balance the equation for the oxidation of $H_2PO_2^-$ to $H_2PO_4^-$ by Cl_2 in acid solution.

1. In writing the skeleton equation, Cl^- is selected as the only possible chlorine-containing product, since Cl_2 is being reduced.

$$H_2PO_2^- + Cl_2 \rightarrow H_2PO_4^- + Cl^-$$

The two half-reactions are

Reduction: $Cl_2 \rightarrow Cl^-$

Oxidation: $H_2PO_2^- \rightarrow H_2PO_4^-$

2.
$$Cl_2 \rightarrow 2Cl^-$$

$$H_2PO_2^- + 2H_2O \rightarrow H_2PO_4^- + 4H^+$$

3.
$$2e^- + Cl_2 \rightarrow 2Cl^-$$
$$H_2PO_2^- + 2H_2O \rightarrow H_2PO_4^- + 4H^+ + 4e^-$$

4. The chlorine half-reaction must be multiplied through by 2:
$$4e^- + 2Cl_2 \rightarrow 4Cl^-$$
$$\underline{H_2PO_2^- + 2H_2O \rightarrow H_2PO_4^- + 4H^+ + 4e^-}$$
$$\cancel{4e^-} + H_2PO_2^- + 2Cl_2 + 2H_2O \rightarrow H_2PO_4^- + 4Cl^- + 4H^+ + \cancel{4e^-}$$

Problems

1. Rewrite each molecular equation, showing as ions each species that dissociates in water. Cancel ions common to each side of the equation. Salts that are insoluble in water are designated (s). (Consult Table 2.2 and the Appendix for K_a values of acids.)

$$HNO_3 + KOH \rightarrow KNO_3 + H_2O$$

$$H_2S + NiCl_2 \rightarrow NiS(s) + 2HCl$$

$$5FeCl_2 + KMnO_4 + 8HCl \rightarrow 5FeCl_3 + MnCl_2 + 4H_2O + KCl$$

2. Write a balanced, net-ionic equation for the reaction indicated.
   a. Hypochlorous acid + sodium hydroxide $\rightarrow$

   b. $NH_3 + HC_2H_3O_2 \rightarrow$

   c. $(NH_4)_2SO_4 + NaOH \rightarrow$

   d. Calcium nitrate + sulfuric acid $\rightarrow$

e. $Cd(NO_3)_2 + 2NaOH \rightarrow$

f. Silver nitrate + magnesium bromide →

g. Barium sulfite + nitric acid →

3. Complete and balance the following oxidation–reduction equations. An underlined blank space indicates that a reasonable choice of a missing principal product must be made. Also, H_2O, H^+, or OH^- must be included where appropriate.

a. $Fe^{2+} + Br_2 \xrightarrow{acid} Fe^{3+} + \underline{}$

b. $O_2 + I^- \xrightarrow{acid} I_2 + H_2O$

c. $ClO^- + Mn(OH)_2(s) \xrightarrow{base} Cl^- + MnO_2$

d. $Zn + NO_3^- \xrightarrow{acid} \underline{} + NH_4^+$

e. $MnO_4^- + H_2O_2 \xrightarrow{acid} Mn^{2+} + O_2$

f. $Al + OH^- \xrightarrow{base} Al(OH)_4^- + \underline{}$

g. $IO_3^- + HSO_3^- \xrightarrow{acid} I_2 + \underline{}$

h. $IO_3^- + I^- \xrightarrow{acid} I_2$

i. $Cr_2O_7^{2-} + CH_3CHO \xrightarrow{acid} \underline{\qquad} + CH_3COOH$

j. $Ag + CN^- + O_2 \xrightarrow{base} Ag(CN)_2^- + H_2O$

4. A net-ionic equation has been written for each reaction below. However, the equations are unsatisfactory because the underlined products are incorrect for the reaction conditions indicated. Rewrite the equations with correct products.

a. HCl is added to a solution containing diamminesilver(I).

$$Ag(NH_3)_2^+ + Cl^- \rightarrow AgCl(s) + \underline{2NH_3}$$

b. Barium carbonate reacts with sulfuric acid.

$$BaCO_3(s) + 2H^+ \rightarrow \underline{Ba^{2+}} + CO_2 + H_2O$$

c. Sodium metal reacts with water.

$$2Na(s) + H_2O \rightarrow H_2 + 2Na^+ + \underline{O^{2-}}$$

d. Sodium thiosulfate solution is treated with Cl_2.

$$S_2O_3^{2-} + 2Cl_2 + 3H_2O \rightarrow \underline{2HSO_3^-} + 4Cl^- + 4H^+$$

e. Chlorate ion is reduced with excess $FeCl_2$ in acid solution.

$$2ClO_3^- + 10Fe^{2+} + 6H^+ \rightarrow \underline{Cl_2} + 10Fe^{3+} + \underline{6OH^-}$$

Equilibrium Calculations

CHAPTER 4

The purpose of Chapter 4 is to establish the techniques for making numerical calculations of equilibrium concentrations in aqueous solution. A brief outline of the important operations is presented in the following sections. More detailed information is readily available in corresponding parts of your general chemistry textbook.

The equilibrium-constant expression for the generalized reaction

$$aA + bB \rightleftarrows cC + dD \tag{1}$$

is

$$K = \frac{[C]^c[D]^d}{[A]^a[B]^b} \tag{2}$$

Each chemical reaction has an equilibrium-constant equation associated with it. Note that these equations are constructed by placing products over reactants. Each concentration term in moles per liter, denoted by brackets, [ ], is raised to a power equal to the stoichiometric coefficient for the corresponding chemical species. By convention, concentration terms for the solvent, H_2O, and for solids are not written in the equilibrium-constant expression. The concentrations of H_2O and of a solid are themselves constants and are included in the value of K. This point can be illustrated by considering the dissociation of H_2O.

Pure water is a molecular substance that is very slightly dissociated into hydrogen ions and hydroxide ions:

$$2H_2O \rightleftarrows H_3O^+ + OH^- \tag{3}$$

$$H_2O \rightleftarrows H^+ + OH^- \tag{4}$$

The equilibrium-constant expression for equation (4) is

$$K_w = [H^+][OH^-] = 1.0 \times 10^{-14} \quad (25°C) \tag{5}$$

The constant, K_w, is called the **ion product** for water. The concentration of H_2O does not appear in equation (5) because H_2O is present in so large an amount that its concentration is essentially constant. The concentration of H_2O, 55.5 M, is included in the value of K_w. In pure water, $[H^+] = [OH^-] = 1.0 \times 10^{-7} M$. However, it is not necessary that H^+ and OH^- concentrations be equal in the presence of added solutes. If, for example, an acid is added, the hydrogen-ion concentration becomes larger than $1.0 \times 10^{-7} M$. At the same time, the concentration of hydroxide ion becomes correspondingly smaller, so the product, $[H^+][OH^-]$, is always equal to 1.0×10^{-14}.

4.1 DISSOCIATION OF WEAK ACIDS AND BASES

The dissociation of a weak, monoprotic acid HA, is shown by

$$HA \rightleftharpoons H^+ + A^- \qquad (6)$$

(To simplify matters, the molecule of H_2O that acts as the proton acceptor has been omitted.) The corresponding equilibrium-constant expression,

$$K_a = \frac{[H^+][A^-]}{[HA]} \qquad (7)$$

contains the acid-dissociation constant, K_a. Acids with very large values of K_a, such as HCl, are completely dissociated in aqueous solution. Acids having very small values of K_a, such as HCN, are only slightly dissociated. In addition to very weak acids of this type, there are a number of moderately weak acids, such as HF and $HC_2H_3O_2$, having intermediate K_a values. K_a values for a number of weak acids and bases are listed in the Appendix.

In describing a solution of a weak acid, it is desirable to know the concentrations of each of the separate species, H^+, A^-, and HA. Using known values of K_a for particular acids, the required concentrations can be obtained. The steps outlined below are useful in finding equilibrium concentrations.

1. Write the equation for dissociation.
2. Write the corresponding equilibrium-constant expression.
3. Define an unknown variable x and express all equilibrium concentrations in terms of initial concentrations and the x variable.
4. Substitute concentrations from step 3 into the equilibrium-constant expression.
5. Solve for x using either (a) the quadratic equation* or (b) the approximation that very little dissociation occurs. Convert the concentration terms to numerical values.

Dissociation of a Weak Acid in Water

Acetic acid frequently is used to provide moderately acidic solutions. The following sample problem illustrates the application of the rules above to calculating $[H^+]$ in an acetic acid solution.

Determine $[H^+]$, $[C_2H_3O_2^-]$, and $[HC_2H_3O_2]$ for a $0.100 M$ $HC_2H_3O_2$ solution. $K_a = 1.8 \times 10^{-5}$ for acetic acid.

*The solution to the quadratic equation $ax^2 + bx + c = 0$ is shown below. The root chosen must lead to positive values of all concentrations.

$$x = \frac{-b \pm \sqrt{b^2 - 4ac}}{2a}$$

1. $$HC_2H_3O_2 \rightleftharpoons H^+ + C_2H_3O_2^-$$

2. $$1.8 \times 10^{-5} = \frac{[H^+][C_2H_3O_2^-]}{[HC_2H_3O_2]}$$

3. Let $x = [H^+]$.

$[H^+] = x$ (by definition)

$[C_2H_3O_2^-] = x$ (H^+ and $C_2H_3O_2^-$ are produced in equal concentrations in step 1; the small concentration of H^+ arising from the dissociation of water may be neglected)

$[HC_2H_3O_2] = 0.100 - x$ (1 mole of $HC_2H_3O_2$ dissociates for every mole of H^+ that appears)

4. $$1.8 \times 10^{-5} = \frac{(x)(x)}{0.100 - x}$$

5. (a) Converting the equation above to quadratic form and solving for x gives $\underline{x = 1.3 \times 10^{-3}\ M}$.

$[H^+] = 1.3 \times 10^{-3}\ M$

$[C_2H_3O_2^-] = 1.3 \times 10^{-3}\ M$

$[HC_2H_3O_2] = 0.100 - 0.0013 = 0.099\ M$

(b) Using the approximation method, assume that, in step 4, $0.100 - x \cong 0.100$ (that is, x is negligibly small compared with 0.100). The equation to be solved is then

$$1.8 \times 10^{-5} = \frac{x^2}{0.100} \quad \text{or} \quad x^2 = 1.8 \times 10^{-6}$$

for which $\underline{x = 1.3 \times 10^{-3}\ M}$, a result identical, using two significant figures, with that in method (a). The assumption that $0.100 - x \cong 0.100$ can be checked. Inserting the value of x gives $0.100 - 0.0013 = 0.099$. Thus, the error introduced by the assumption is about 1 percent. For present purposes, errors in the range 1 to 3 percent are acceptable. The rigorously correct answer can always be obtained using the quadratic equation. However, method (b) is much simpler to use and gives reliable results when K_a values are 10^{-5}, or smaller. For acids with larger K_a values, method (a), or a successive application of method (b), must be used.

From the results above, it is possible to compute the **percent dissociation** of the acetic acid.

$$\text{percent dissociation} = \frac{[H^+]}{0.100} \times 100 = \frac{(1.3 \times 10^{-3})(100)}{0.100} = \underline{1.3}\text{ percent}$$

In calculating the percent dissociation, the concentration of H^+ produced is divided by the *initial* concentration of $HC_2H_3O_2$. It is also possible to determine the equilibrium concentration of OH^- present, using equation (5):

$$[OH^-] = \frac{1.0 \times 10^{-14}}{[H^+]} = \frac{1.0 \times 10^{-14}}{1.3 \times 10^{-3}} = \underline{7.7 \times 10^{-12} M}$$

Calculations for the dissociation of a weak base are handled as described above. It is important that steps 1 and 2 be written out properly. Using the weak base ammonia, NH_3, as an example, steps 1 and 2 are written as follows:

1.
$$NH_3 + H_2O \rightleftarrows NH_4^+ + OH^-$$

2.

$$K_b = \frac{[NH_4^+][OH^-]}{[NH_3]}$$

Dissociation of a Weak Acid in a Buffer Solution

In the acid-dissociation calculations above, it was permissible to set $[H^+] = [C_2H_3O_2^-]$, since the only significant source of these ions was dissociation of $HC_2H_3O_2$. However, there are many solutions in which a salt of the weak acid is also present, so $[H^+]$ is not equal to $[A^-]$ but is much less than $[A^-]$. Two ways of preparing such solutions are to mix a solution of HA with NaA, or to partially neutralize a solution of HA with NaOH, as shown by

$$HA + NaOH \rightarrow NaA + H_2O \quad (8)$$

After partial neutralization, the solution contains both NaA and unreacted HA. Solutions containing a weak acid and the salt of that weak acid (or a weak base and the salt of that weak base) are called **buffer solutions**. The hydrogen-ion concentration in a buffer solution can be calculated using steps 1 through 5 outlined previously.

What is $[H^+]$ in a solution that contains $0.10\ M\ HC_2H_3O_2$ and $0.20\ M\ NaC_2H_3O_2$? $K_a = 1.8 \times 10^{-5}$ for $HC_2H_3O_2$.

1. $HC_2H_3O_2 \rightleftarrows H^+ + C_2H_3O_2^-$. (This is the only significant dissociation equilibrium involved. $NaC_2H_3O_2$ is 100 percent dissociated, so dissociation of $NaC_2H_3O_2$ is *not* the proper equilibrium to use.)

2.

$$1.8 \times 10^{-5} = \frac{[H^+][C_2H_3O_2^-]}{[HC_2H_3O_2]}$$

3. Let $x = [H^+]$.

$[H^+] = x$ (by definition)

$[C_2H_3O_2^-] = 0.20 + x$ (the initial concentration of $C_2H_3O_2^-$, plus that which arises from dissociation of $HC_2H_3O_2$)

$[HC_2H_3O_2] = 0.10 - x$ (1 mole of $HC_2H_3O_2$ dissociates for every mole of H^+ that appears)

4.
$$1.8 \times 10^{-5} = \frac{(x)(0.20 + x)}{0.10 - x}$$

5. (a) Solving the quadratic equation, $\underline{x = 9.0 \times 10^{-6} \, M}$.

$[H^+] = 9.0 \times 10^{-6} \, M$

$[C_2H_3O_2^-] = 0.20 + (9.0 \times 10^{-6}) = 0.20$

$[HC_2H_3O_2] = 0.10 - (9.0 \times 10^{-6}) = 0.10$

(b) Using the approximation method, assume that in step 4, $0.20 + x \cong 0.20$ and that $0.10 - x \cong 0.10$. The equation to be solved is

$$1.8 \times 10^{-5} = \frac{(x)(0.20)}{0.10}$$

for which $\underline{x = 9.0 \times 10^{-6} \, M}$, a result identical with that in (a).

Method (b) is almost always satisfactory in buffer solutions, since the effect of the added salt is to repress the dissociation of the weak acid. The percent dissociation is $(9.0 \times 10^{-6}) \times (100)/0.10 = \underline{9.0 \times 10^{-3} \text{ percent}}$, which is much less than the percent dissociation observed for $HC_2H_3O_2$ in the absence of $NaC_2H_3O_2$.

Buffer solutions possess two characteristics that make them valuable in the cation and anion experiments. The acidity of a solution may be adjusted as desired by varying the ratio of $[A^-]/[HA]$. Buffer solutions also resist changes in acidity. Added OH^- reacts with and is neutralized by the HA present, and added H^+ reacts with and is tied up by the A^- present. These same two characteristics are also responsible for the great importance of buffer solutions in the many biological systems in which they occur.

4.2 ⎰ pH. MEASURING ACIDITY IN SOLUTION

pH

Solutions commonly encountered in the laboratory range in hydrogen-ion concentration from 1 to $10^{-14} \, M$. The pH scale is often used to describe solutions that vary greatly in acidity. In dilute solutions, pH is defined as follows:

$$\text{pH} = -\log_{10} [H^+] \quad \text{or} \quad [H^+] = 10^{-\text{pH}}$$

Thus, pH is a numerical exponent, the power to which the base 10 is raised in order to correctly express $[H^+]$.

Integral pH values. pH is an integer for any solution in which $[H^+]$ is an exact power of 10.

$[H^+]$	1.0	1.0×10^{-1}	1.0×10^{-7}	1.0×10^{-13}
pH	0	1	7	13

In response to the question, What is the pH of 1.0×10^{-2} M HCl?, an additional piece of information is required. Since HCl is 100 percent dissociated, the solution contains $[H^+] = [Cl^-] = 1.0 \times 10^{-2}$ M. Thus, pH = 2.

Nonintegral pH values. For solutions in which $[H^+]$ is not an exact power of 10, logarithms are required to calculate pH.

$[H^+]$	2.0×10^{-4}	5.0×10^{-8}	8.3×10^{-12}
pH	3.7	7.3	11.1

The pH corresponding to $[H^+] = 2.0 \times 10^{-4}$ is calculated using the relationship $10^a \times 10^b = 10^{a+b}$. Using the slide rule or a log table, the log of 2.0 is 0.30. This means that 2.0 may also be written as $10^{0.30}$.

$$[H^+] = 2.0 \times 10^{-4} = 10^{0.30} \times 10^{-4} = 10^{0.30-4} = 10^{-3.7}$$

or

$$pH = 3.7$$

Note that by using the sequence of equalities in the line above, it is possible to answer the question: What is $[H^+]$ in a solution that has pH = 3.7?

The useful range of the pH scale, shown below, extends from pH = 1 in highly acid solution to pH = 13 in highly basic solution. Note that a *low* pH value corresponds to a *high* $[H^+]$.

pH = 1	3	5	7	9	11	13
$[H^+] = 0.1$			$[H^+] = [OH^-]$ $= 10^{-7}$			$[OH^-] = 0.1$
very acidic	*moderately acidic*		*neutral*		*moderately basic*	*very basic*

pH Indicators

pH can be measured precisely with an electronic voltmeter, called a pH meter. However, for the experimental work in Parts II and III, it is preferable to use a very rapid means of qualitatively estimating pH. Wide-range pH indicator paper is ideal for this purpose. When a drop of solution to be tested is placed onto this paper, the paper immediately assumes a color that indicates the pH range of the solution. The paper contains a mixture of indicator molecules which are weak acids. A particular indicator may be represented as HIn, which dissociates in solution:

$$HIn \rightleftarrows H^+ + In^- \qquad (9)$$

The corresponding equilibrium-constant expression involves a dissociation constant, K_{HIn}:

$$K_{HIn} = [H^+]\frac{[In^-]}{[HIn]} \qquad (10)$$

Usually, both the molecular form HIn, and the anion form, In^-, of the indicator have intense, but different colors. Thus, the effective color of an indicator mixture depends on the ratio of $[In^-]/[HIn]$. As seen in equation (10), this ratio is dependent on $[H^+]$ in solution. By properly combining indicators, it is possible to get a continuous change in color as $[H^+]$ is changed over a wide range of values. Several indicators are listed below, along with the colors of the HIn and In^- forms, and the pH range in which HIn and In^- are both present in large enough concentrations to give a mixed color.

Indicator	pH Range	HIn	In⁻
Methyl orange	3.1–4.4	Red	Yellow
Bromthymol blue	6.0–7.6	Yellow	Blue
Alizarin yellow	10.1–12.0	Yellow	Violet

4.3　SOLUBILITY OF SALTS

Equilibrium systems involving salts of low solubility have been discussed in Section 2.3. In the present section the quantitative aspects of solubility are considered. Using the techniques outlined in steps 1 through 5 in Section 4.1, it should be possible to calculate equilibrium concentrations of cations and anions in solution.

Calculation of Equilibrium Concentrations

As relevant examples of equilibrium calculations, let us consider the solubility of two salts that are encountered in the first experiments in Part II. Silver chloride, AgCl, and lead chloride, $PbCl_2$, are both salts of low solubility in water. Systems involving excess solid in equilibrium with the saturated solution are represented as follows:

$$AgCl(s) \rightleftarrows Ag^+ + Cl^- \qquad (11)$$

$$PbCl_2(s) \rightleftarrows Pb^{2+} + 2Cl^- \qquad (12)$$

In writing the corresponding equilibrium-constant expressions,

$$K_{sp} = [Ag^+][Cl^-] \qquad (13)$$

$$K_{sp} = [Pb^{2+}][Cl^-]^2 \qquad (14)$$

the concentrations of the solids are not written, since these values are constant. The equilibrium constants are symbolized K_{sp} and are called

solubility products because they are simply products of concentrations of ions resulting from dissolving of the salt. The K_{sp} values vary over a wide range. For example, at 25°C, $K_{sp} = 1.6 \times 10^{-5}$ for $PbCl_2$, $K_{sp} = 1.6 \times 10^{-10}$ for $AgCl$, $K_{sp} = 8 \times 10^{-28}$ for PbS, and $K_{sp} = 1 \times 10^{-51}$ for Ag_2S. It is important to remember that K_{sp} values and other equilibrium constants change considerably with temperature. K_{sp} values usually increase with temperature and also with increasing concentration of other ions in solution (an ion-activity effect). These K_{sp} values can be used in conjunction with steps 1 through 5 in Section 4.1 to calculate equilibrium concentrations.

Determine $[Ag^+]$ and $[Cl^-]$ in equilibrium with solid AgCl in water at 25°C. $K_{sp} = 1.6 \times 10^{-10}$ for AgCl.

1. $$AgCl(s) \rightleftarrows Ag^+ + Cl^-$$

2. $$1.6 \times 10^{-10} = [Ag^+][Cl^-]$$

3. Let $[Ag^+] = s$.

$[Ag^+] = s$ ($[Ag^+]$ is equal to the *solubility, s,* of AgCl in water.)

$[Cl^-] = s$ ($[Ag^+]$ and $[Cl^-]$ are produced in equal concentrations as AgCl dissolves in water.)

4. $$1.6 \times 10^{-10} = (s)(s) = s^2$$

5. The above equation can be solved by taking the square root of both sides:

$s^2 = 1.6 \times 10^{-10}$ or $\underline{s = [Ag^+] = [Cl^-] = 1.3 \times 10^{-5} \, M}$

Note that this process can be reversed in order to calculate K_{sp}, given a value of the observed solubility, s, since $K_{sp} = s^2$ for AgCl.

Determine $[Pb^{2+}]$ and $[Cl^-]$ in equilibrium with solid $PbCl_2$ in water. $K_{sp} = 1.6 \times 10^{-5}$ for $PbCl_2$.

1. $$PbCl_2(s) \rightleftarrows Pb^{2+} + 2Cl^-$$

2. $$1.6 \times 10^{-5} = [Pb^{2+}][Cl^-]^2$$

3. Let $[Pb^{2+}] = s$.

$[Pb^{2+}] = s$ ($[Pb^{2+}]$ is equal to the *solubility, s,* of $PbCl_2$ in water)

$[Cl^-] = 2s$ (two Cl^- are produced for each Pb^{2+} that appears)

4. $1.6 \times 10^{-5} = (s)(2s)^2 = 4s^3$ or $s^3 = 4.0 \times 10^{-6}$

5. The equation in step 4 can be solved taking the cube root of both sides. (Note that the exponent is divided by 3.)

$s = 1.6 \times 10^{-2}$, $\underline{[Pb^{2+}] = 1.6 \times 10^{-2}\, M}$,

$\underline{[Cl^-] = 3.2 \times 10^{-2}\, M}$

In the sample problems above it was assumed that AgCl and PbCl$_2$ were dissolved in initially pure water. However, suppose that solid AgCl is added to a solution that already contains 0.20 M Cl$^-$ from added 0.20 M HCl. Equilibrium reaction (11) and equilibrium-constant equation (13) will still hold. However, applying Le Châtelier's principle, the solubility of AgCl in the presence of Cl$^-$ should decrease. This is an example of the **common-ion effect.**

Determine [Ag$^+$] in a 0.20 M HCl solution that has been saturated with AgCl.

1. $AgCl(s) \rightleftarrows Ag^+ + Cl^-$

2. $1.6 \times 10^{-10} = [Ag^+][Cl^-]$

3. Let [Ag$^+$] = s.

[Ag$^+$] = s ([Ag$^+$] is a measure of the concentration of AgCl that dissolves)

[Cl$^-$] = 0.20 + s ([Cl$^-$] is equal to the concentration of Cl$^-$ initially present in solution plus that due to dissolving of AgCl)

4. $1.6 \times 10^{-10} = (s)(0.20 + s)$

5. (**b**) It is unnecessary to use the quadratic equation in step 4, since it is known that s should be very small compared with 0.20. Assume that $0.20 + s \cong 0.20$. Then

$1.6 \times 10^{-10} = (s)(0.20)$ or $\underline{s = [Ag^+] = 8.0 \times 10^{-10}\, M}$

The assumption that s may be neglected in comparison with 0.20 is a very good one. Note that the presence of added chloride ion has decreased the equilibrium concentration of silver ion to $8.0 \times 10^{-10}\, M$, compared with a value of $1.3 \times 10^{-5}\, M$ in water.

Extent of Precipitation Using K_{sp}

An important operation in the cation chemistry in Part II is the complete precipitation of a metal ion. This is accomplished by adding an excess of the precipitating anion. Values of K_{sp} can be used to predict, first, whether or not precipitation will occur when anion and cation solutions are mixed, and, second, the extent to which precipitation will occur. The solubility product of the ion concentrations provides the necessary criterion. If, on mixing the cation and anion, the appropriate product of ion concentrations is less than the value of K_{sp}, no precipitation will occur. However, if the product of ion concentrations exceeds the value of K_{sp}, precipitation must occur. Precipitation will take place until the concentrations decrease to the point at which the product of ion concentrations is just equal to K_{sp}. The system of solid and saturated solution will then be in equilibrium. These ideas can be illustrated by considering precipitation of Pb^{2+} from a $Pb(NO_3)_2$ solution by addition of HCl:

$$Pb^{2+} + 2Cl^- \rightarrow PbCl_2(s) \tag{15}$$

This precipitation reaction is used in the first experiment in Part II to remove lead ion from solution. The lead ion concentration encountered there will be approximately 0.030 M.

Predict whether or not precipitation of $PbCl_2$ will occur if, to 0.030 M Pb^{2+}, chloride ion is added at a 0.010 M concentration level. $K_{sp} = 1.6 \times 10^{-5}$.

The K_{sp} expression for $PbCl_2$ is $K_{sp} = [Pb^{2+}][Cl^-]^2$. The product of ion concentrations must be compared with the K_{sp} value.

$$[Pb^{2+}][Cl^-]^2 = (0.030)(0.010)^2 = \underline{3.0 \times 10^{-6} < 1.6 \times 10^{-5}}$$

Precipitation will not occur because the ion product is less than the value of K_{sp}. Insufficient chloride has been added, so the solution does not reach the saturation level.

Predict whether or not precipitation of $PbCl_2$ will occur if, to 0.030 M Pb^{2+}, chloride ion is added to give a concentration of 1.0 M. If precipitation occurs, calculate the equilibrium concentration of Pb^{2+} remaining in solution. (Assume that $[Cl^-]$ remains equal to 1.0 M.) $K_{sp} = 1.6 \times 10^{-5}$ for $PbCl_2$.

The product of ion concentrations is

$$[Pb^{2+}][Cl^-]^2 = (0.030)(1.0)^2 = \underline{3.0 \times 10^{-2} \gg 1.6 \times 10^{-5}}$$

so precipitation does occur. Since chloride ion is initially added in large excess, its concentration remains essentially equal to 1.0 M. Equation (14) is valid since $PbCl_2$ is present following precipitation.

1. $$PbCl_2(s) \rightleftarrows Pb^{2+} + 2Cl^-$$

2. $$1.6 \times 10^{-5} = [Pb^{2+}][Cl^-]^2$$

3. Let $x = [\text{Pb}^{2+}]$.

$[\text{Pb}^{2+}] = x$ (by definition)

$[\text{Cl}^-] = 1.0\ M$ (Cl$^-$ is present in large excess; a more precise value would be $1.0 - 2x$)

4. $$1.6 \times 10^{-5} = (x)(1.0)^2$$

5. From step 4, $\underline{x = [\text{Pb}^{2+}] = 1.6 \times 10^{-5}\ M}$.

Thus, lead ion precipitates until its concentration falls from the initial value of $0.030\ M$ to a final concentration of $1.6 \times 10^{-5}\ M$. This corresponds to nearly complete precipitation, since only $(1.6 \times 10^{-5})(100)/0.030 = 5$ percent of the original Pb^{2+} remains in solution. Actually, PbCl$_2$ is only moderately insoluble in cold water. Most other precipitates that will be encountered have much smaller values of K_{sp} and, therefore, the metal ion involved is very effectively removed from solution.

4.4 SIMULTANEOUS EQUILIBRIA. PRECIPITATION OF SULFIDES

Dissociation of H$_2$S

In water solution hydrogen sulfide, H$_2$S, is a weak, diprotic acid. The two protons dissociate in stepwise fashion, the first proton being more easily lost than the second. The two acid-dissociation constants are written as K_1 and K_2:

$$\text{H}_2\text{S} \rightleftarrows \text{H}^+ + \text{HS}^- \tag{16}$$

$$\text{HS}^- \rightleftarrows \text{H}^+ + \text{S}^{2-} \tag{17}$$

$$K_1 = 1 \times 10^{-7} = \frac{[\text{H}^+][\text{HS}^-]}{[\text{H}_2\text{S}]} \tag{18}$$

$$K_2 = 1 \times 10^{-14} = \frac{[\text{H}^+][\text{S}^{2-}]}{[\text{HS}^-]} \tag{19}$$

These equilibria are of importance because sulfide ion, S^{2-}, is used as the precipitating anion in a number of the experimental steps in Part II. Since K_1 is quite small, only very low concentrations of HS$^-$ are produced. This ion, in turn, dissociates very slightly to give S^{2-} ions in the K_2 equilibrium step. In fact, it would appear that solutions of H$_2$S provide too small a concentration of S^{2-} to be of any use in precipitating metal ions. However, a number of metal ions form exceedingly insoluble sulfides, so H$_2$S can be effective in precipitating the metal sulfides. In general, the reaction for dissolving of divalent metal sulfides is

$$\text{MS(s)} \rightleftarrows \text{M}^{2+} + \text{S}^{2-} \tag{20}$$

for which the equilibrium-constant expression is

$$K_{sp} = [\text{M}^{2+}][\text{S}^{2-}] \tag{21}$$

When solutions of a metal ion, M^{2+}, and H_2S are mixed, reaction (20) and the H_2S dissociation reactions occur simultaneously. The equilibrium systems are linked together by the sulfide ion. Thus, S^{2-} must be simultaneously in equilibrium with both H^+ and M^{2+}.

Metal sulfide precipitations are carried out in the presence of a known, excess hydrogen-ion concentration in solutions saturated with H_2S. In these saturated solutions, the concentration of H_2S remains constant at 0.10 M. Under these conditions, equations (18) and (19) can be combined by multiplication to give

$$K_1 K_2 = \frac{[H^+][HS^-]}{[H_2S]} \times \frac{[H^+][S^{2-}]}{[HS^-]} = \frac{[H^+]^2[S^{2-}]}{[H_2S]} \qquad (22)$$

Inserting values of the two constants and the value of $[H_2S] = 0.10$ results in the simplified equation

$$(1 \times 10^{-7})(1 \times 10^{-14}) = \frac{[H^+]^2[S^{2-}]}{0.10}$$

or

$$[H^+]^2[S^{2+}] = 1 \times 10^{-22} \qquad (23)$$

Equation (23) is valid for saturated H_2S solutions at 25°C with hydrogen-ion concentrations set by some external source of H^+ such as added HCl or an appropriate buffer solution.

Selective Precipitation of Metal Sulfides

A practical problem that arises in the experiments in Part II is the selective precipitation of one metal ion, leaving the other ions in solution. This provides an effective means of separating metal ions. Suppose that we have a solution that contains a mixture of 0.10 M $Ni(NO_3)_2$ and 0.10 M $Cd(NO_3)_2$. The two metal ions, Ni^{2+} and Cd^{2+}, form the insoluble sulfides, NiS and CdS:

$$NiS(s) \rightleftarrows Ni^{2+} + S^{2-} \qquad (24)$$

$$K_{sp} = 2 \times 10^{-21} = [Ni^{2+}][S^{2-}] \qquad (25)$$

$$CdS(s) \rightleftarrows Cd^{2+} + S^{2-} \qquad (26)$$

$$K_{sp} = 8 \times 10^{-27} = [Cd^{2+}][S^{2-}] \qquad (27)$$

Since CdS is the more insoluble of the two sulfides, it should be possible to devise conditions such that Cd^{2+} can be precipitated by H_2S while leaving Ni^{2+} in solution. This can be accomplished by adjusting $[S^{2-}]$ to an appropriate level. Equation (23) suggests a means of controlling $[S^{2-}]$ since, as $[H^+]$ increases, $[S^{2-}]$ must decrease in a saturated H_2S solution. A typical experimental situation is described in the following sample problems.

A solution containing 0.3 M H^+ is saturated with H_2S at 25°C. What will be the equilibrium concentration of sulfide ion?

Using equation (23), $[H^+]^2[S^{2-}] = 1 \times 10^{-22}$,

$(0.3)^2[S^{2-}] = 1 \times 10^{-22}$ or $\underline{[S^{2-}] = 1 \times 10^{-21} M}$

A solution containing $0.10\ M$ $Ni(NO_3)_2$ and $0.10\ M$ $Cd(NO_3)_2$ is acidified so that $[H^+] = 0.3\ M$. The mixture is then saturated with H_2S. Will either NiS or CdS precipitate? Calculate the equilibrium concentrations of Ni^{2+} and Cd^{2+} in this system.

To predict whether or not precipitation occurs, it is necessary to compute the appropriate product of ion concentrations for nickel sulfide and cadmium sulfide. The equilibrium concentration of sulfide ion, $[S^{2-}] = 1 \times 10^{-21} M$, can be used from the preceding problem. $K_{sp} = 2 \times 10^{-21}$ for NiS. $K_{sp} = 8 \times 10^{-27}$ for CdS.

$[Ni^{2+}][S^{2-}] = (0.10)(1 \times 10^{-21}) = \underline{1 \times 10^{-22} < 2 \times 10^{-21}}$

The product of ion concentrations does not exceed K_{sp} for NiS; therefore, NiS will not precipitate. $[Ni^{2+}]$ remains equal to $0.10\ M$.

$[Cd^{2+}][S^{2-}] = (0.10)(1 \times 10^{-21}) = \underline{1 \times 10^{-22} \gg 8 \times 10^{-27}}$

CdS will precipitate because the product of ion concentrations exceeds K_{sp}. Cd^{2+} is converted to CdS until $[Cd^{2+}]$ is in equilibrium with sulfide ion.

At equilibrium, $[Cd^{2+}][S^{2-}] = 8 \times 10^{-27}$, or

$[Cd^{2+}](1 \times 10^{-21}) = 8 \times 10^{-27}$, $\underline{[Cd^{2+}] = 8 \times 10^{-6} M}$

Thus, Ni^{2+} remains in solution but Cd^{2+} is essentially 100 percent precipitated as CdS. Sulfide-ion precipitation results in quantitative separation of the two metal ions.

Problems

1. Write the equilibrium-constant expression that corresponds to each of the following reactions.

  a. $HNO_2 \rightleftarrows H^+ + NO_2^-$

  b. $CO_3^{2-} + H_2O \rightleftarrows HCO_3^- + OH^-$

44 Equilibrium Calculations

    c. $Ag_2CrO_4(s) \rightleftarrows 2Ag^+ + CrO_4^{2-}$

    d. $Cd(NH_3)_4^{2+} \rightleftarrows Cd^{2+} + 4NH_3$

    e. $N_2(g) + 3H_2(g) \rightleftarrows 2NH_3(g)$

    f. $CaCO_3(s) \rightleftarrows CaO(s) + CO_2(g)$

2. Hypochlorous acid dissociates slightly in water to given H^+ and ClO^-. Using the value of $K_a = 3.0 \times 10^{-8}$ for HClO, calculate the hydrogen-ion concentration in a 0.10 M HClO solution. What is the percent dissociation of HClO?

3. Calculate $[OH^-]$ in a 0.50 M NH_3 solution in water. $K_b = 1.8 \times 10^{-5}$ at 25°C for NH_3.

4. A mixture of NH_3 and its salt, NH_4Cl, constitutes a buffer system. Ammonia buffer solutions are used in several of the experimental procedures in Part II. Calculate the hydroxide-ion concentration in a buffer mixture containing 0.20 M NH_3 and 0.50 M NH_4Cl. What is $[H^+]$ in this solution? $K_b = 1.8 \times 10^{-5}$ for NH_3.

5. It is desired to prepare a buffer solution in which $[H^+] = 1.0 \times 10^{-5}\ M$. Acetic acid is selected as the weak acid because its K_a value of 1.8×10^{-5} is close to the required hydrogen-ion concentration. In a solution containing 0.60 M $HC_2H_3O_2$, what concentration of $C_2H_3O_2^-$, obtained by addition of $NaC_2H_3O_2$, is required to make $[H^+] = 1.0 \times 10^{-5}\ M$?

6. Suggest one buffer system that could be employed to provide solutions in which the hydrogen-ion concentration is held constant at (a) $1.0 \times 10^{-7}\ M$, (b) $1.0 \times 10^{-2}\ M$. Answer by writing, in each case, the formulas of the two buffer components necessary. (Consult K_a values in the Appendix.)

7. Determine the value of pH in solutions containing

    a. $[H^+] = 1.0 \times 10^{-6}\ M$.   *b.* 0.30 M HCl.

    c. 0.050 M NaOH.

Equilibrium Calculations

8. Determine [H$^+$] in a solution in which
   a. pH = 9.0.   b. pH = 3.60.

9. Determine the color of a solution containing the indicator bromthymol blue if the pH were
   a. 5.0_____   b. 6.8_____   c. 8.0_____

10. Determine the solubility of AgI in water, assuming that the only soluble species present are Ag$^+$ and I$^-$. $K_{sp} = 1.5 \times 10^{-16}$ for AgI.

11. At 25°C, $K_{sp} = 4 \times 10^{-11}$ for solutions of CaF$_2$ in water.
    a. Calculate [Ca^{2+}] and [F$^-$] in water that has been saturated with calcium fluoride.

    b. Calculate [Ca^{2+}] in a solution prepared by saturating 0.25 M NaF with CaF$_2$.

12. To a solution containing 0.020 M Ca^{2+} is added Na$_2$CrO$_4$ until [CrO$_4^{2-}$] = 3.0 × 10^{-2} M.
    a. Show whether or not precipitation of calcium chromate will occur in this mixture. $K_{sp} = 7 \times 10^{-4}$ for CaCrO$_4$.

    b. What concentration of CrO$_4^{2-}$ is required to cause precipitation of CaCrO$_4$?

13. Hydrochloric acid is added to a 0.1 M solution of $Cu(NO_3)_2$ until $[H^+] = 0.3\ M$. The solution is saturated with H_2S, resulting in the precipitation of CuS. Calculate the equilibrium concentration of Cu^{2+}. $K_{sp} = 8 \times 10^{-45}$ for CuS.

14. What hydrogen-ion concentration is necessary to just cause precipitation of FeS from a 0.1 M solution of $FeCl_2$ that has been saturated with H_2S? $K_{sp} = 4 \times 10^{-19}$ for FeS.

Coordination Compounds of Transition Metals

CHAPTER 5

5.1 COMPLEX IONS

Central Metal Ion

A complex ion is a chemical species consisting of a metal ion bonded to one or more anions or neutral molecules called ligands. Chemical compounds that contain one or more complex ions are referred to as **coordination compounds**. The most extensively characterized complex ions involve transition metal cations. Although more than one metal ion may be involved, only complex ions with a single metal ion will be considered. Some of the common cations that form complex ions are Ag^+, Cu^{2+}, Pt^{2+}, Fe^{2+}, Fe^{3+}, Cr^{3+}, Co^{3+}, Ni^{2+}, and Al^{3+}. Metals in the 0 oxidation state also form coordination compounds: for example, $Fe(CO)_5$. However, such compounds are seldom encountered in the aqueous chemistry of transition metals.

Using Cr^{3+} as a representative example, several complex ions and the corresponding coordination compounds are shown as follows:

Complex Ion	Coordination Compound
$Cr(H_2O)_6^{3+}$	$[Cr(H_2O)_6]Cl_3$
$Cr(NH_3)_5Cl^{2+}$	$[Cr(NH_3)_5Cl]Cl_2$
$Cr(NH_3)_2(NCS)_4^-$	$K[Cr(NH_3)_2(NCS)_4]$

In each coordination compound above, the complex ion is enclosed in brackets. In the compound $[Cr(NH_3)_5Cl]Cl_2$, the $Cr(NH_3)_5Cl^{2+}$ species is present as a discrete unit. The five NH_3 molecules and Cl^- ion are said to be in the first coordination sphere of Cr^{3+}. The remaining ions, such as the two Cl^- ions in $[Cr(NH_3)_5Cl]Cl_2$ or the K^+ ion in $K[Cr(NH_3)_2(NCS)_4]$, are present in the solid to preserve electroneutrality. Many coordination compounds are soluble in water and, like other ionic compounds, are usually dissociated into ions. Thus, in water, $[Cr(NH_3)_5Cl]Cl_2$ is present as the complex ion $Cr(NH_3)_5Cl^{2+}$ and two Cl^- ions.

Because of the great variety of complex ions that can be formed, the study of coordination chemistry has become one of the most interesting and fastest-moving areas of chemistry. Coordination compounds are especially important in biological systems and in the chemical production of synthetic materials. Complex ions exhibit many different geometries, spectral and magnetic properties, and a variety of bonding situations. In the remainder of this chapter, and in Chapter 6, these important aspects of complex ions will be considered.

Ligands

Molecules or anions that are bonded to the central metal ion are called **ligands**. To be a ligand, a molecule or anion must possess one or more unshared pairs of electrons. Thus, all the anions listed in Table 1.1 could serve as ligands. The NH_3 molecule with its unshared pair of electrons is a common ligand, but, by way of contrast, CH_4 cannot serve as a ligand, since all its electrons are involved in internal covalent bonds. Only one atom in a polyatomic ligand can be bonded to a given coordination site of the metal ion. This atom, which must contain one or more unshared pairs of electrons, is termed the **donor atom**. Some ligands can have two, or more, donor atoms, each of which is bonded to a separate coordination site of the metal ion. Such ligands are referred to as **chelating ligands** and the resulting complex ions as **chelates**.

Table 5.1 | *Some Common Ligands*

$:\!\ddot{O}\!-\!H$, H	aquo	H, $:\!N\!-\!H$, H	ammine
$:\!\ddot{F}\!:^{-}$	fluoro	$:\!\ddot{C}l\!:^{-}$	chloro
$:\!\ddot{B}r\!:^{-}$	bromo	$:\!\ddot{I}\!:^{-}$	iodo
$:\!\ddot{O}H^{-}$	hydroxo	$:\!C\!\equiv\!N\!:^{-}$	cyano
$:\!NCS\!:^{-}$	thiocyanato	$:\!\ddot{S}\!:^{2-}$	thio
$:\!N\!-\!\ddot{O}\!:^{(-)}$, $\|\|$, $:\!\ddot{O}\!:$	nitro	$:\!\ddot{O}\!:^{(2-)}$, $:\!\ddot{S}\!-\!S\!-\!\ddot{O}\!:$, $:\!\ddot{O}\!:$	thiosulfato
$:\!\ddot{O}\!:$ $\ddot{O}\!:^{(2-)}$ $\diagdown\!/$ $C\!-\!C$ $\diagup\!\diagdown$ $\ddot{O}\!:$ $\ddot{O}\!:$	oxalato	$H\!-\!\underset{\|}{N}\!-\!CH_2CH_2\!-\!\underset{\|}{N}\!-\!H$, $\ \ \ H\ \ \ \ \ \ \ \ \ \ \ \ H$	ethylenediamine (en)

The names and structural formulas of a number of ligands are listed in Table 5.1. For polyatomic ligands, the atom that is usually the donor atom is shown in boldface type. Note that the anionic ligands are named using the characteristic name of the anion and the suffix, -o. The ethylenediamine molecule (often abbreviated *en*) and the oxalate anion are examples of chelating ligands. Each of these ligands has two donor atoms and is often referred to as a **bidentate ligand**. When bonded to the metal ion, the ligands in Table 5.1 retain their basic three-dimensional shape. For example, the NH_3 ligand is pyramidal in shape, and the $S_2O_3^{2-}$ ligand possess a nearly tetrahedral geometry.

5.2 { NAMING COORDINATION COMPOUNDS

A coordination compound usually consists of one complex ion and one or more regular ions: for example, K^+, NH_4^+, Cl^-, and NO_3^-. The naming of a complex ion involves a systematic identification of the central metal ion and the number and kind of each of the ligands present. Rules for naming coordination compounds are summarized as follows:

1. Name the cation first, then the anion.
2. In the complex ion, name the ligands first, then the central metal ion.
   (a) Anionic ligands are named first, using the ligand names with the *-o* ending shown in Table 5.1. If more than one type of anionic ligand is present, the ligands should be named in alphabetical order.
   (b) Neutral ligands are named second. Neutral ligands are named in alphabetical order using the ordinary name of the molecule. However, the two common ligands, H_2O, named **aquo**, and NH_3, named **ammine**, are listed first, and in the order H_2O before NH_3.
   (c) To indicate the number of ligands of each type, use the Greek prefixes mono-(1), di-(2), tri-(3), tetra-(4), penta-(5), and hexa-(6).
3. The oxidation state of the central metal ion is indicated with a Roman numeral in parentheses, immediately following the name of the metal. For negative complex ions, the suffix *-ate* is placed after the name of the metal ion.
4. When writing the formula corresponding to a particular name, reverse the order in rule 2; that is, the neutral molecules should be placed immediately following the metal ion:

$[Cr(H_2O)_6](NO_3)_3$	Hexaaquochromium(III) nitrate
$[Co(NH_3)_4Cl_2]Cl$	Dichlorotetraaminecobalt(III) chloride
$[Fe(H_2O)_5NCS]^{2+}$	Thiocyanatopentaaquoiron(III) ion
$Cr(NH_3)_3Cl_3$	Trichlorotriamminechromium(III)
$K[AgCl_2]$	Potassium dichloroargentate(I)
$K_4[Fe(CN)_6]$	Potassium hexacyanoferrate(II)
$Fe(CN)_6^{3-}$	Hexacyanoferrate(III) ion (common name—ferricyanide ion)
$[Pt(NH_3)_4][PtCl_4]$	Tetraammineplatinum(II) tetrachloroplatinate(II)

5.3 COORDINATION NUMBERS AND STRUCTURES OF COMPLEX IONS

Metal ions are bonded directly to a specific number of ligands which comprise the first coordination sphere. The number of monodentate ligands bound, or in the case of multidentate ligands, the number of donor atoms bound is called the **coordination number** (C.N.) of the metal ion. Transition metal ions have characteristic coordination numbers. Usually all the available coordination sites are occupied, although the coordination number of a particular cation may be lower when large groups are present. Bulky ligands cause crowding in the first coordination sphere, a factor that is called a **steric effect**. Associated with each coordination number are one or more characteristic structures. Examples of coordination numbers and structures for typical metal ions are shown in Table 5.2. In Table 5.2. the larger spheres represent the central metal ion and the smaller spheres represent the ligands. The angle between two adjacent ligands and the central ion is indicated in parentheses.

Table 5.2 *Coordination Numbers and Structures*

Metal ion	C.N.	Structure
Ag^+	2	Linear (180°)
Hg^{2+}, Zn^{2+}	4	Tetrahedral (109°)
Pt^{2+}, Ni^{2+}	4	Square planar (90°)
Cu^{2+}, Ni^{2+}, Pt^{4+}, Fe^{2+}, Cr^{3+}, Fe^{3+}, Co^{3+}, Al^{3+}	6	Octahedral (90°)

Each of the three-dimensional structures is highly symmetrical, so in a given structure all the ligands occupy equivalent positions. Using x-ray diffraction techniques, the structures of a large number of complex ions have recently been determined.

Most transition metal ions can form complexes of more than one structural type. The coordination numbers in Table 5.2 are those often observed in the more common complexes in aqueous solution. Two other points should be noted in conjunction with Table 5.2. First, coordination numbers increase, in general, with an increase in oxidation number of the metal ion. Second, although coordination numbers of 3, 5, and 7 are known, the even-numbered values of C.N. are far more frequently observed.

Complex ions with various compositions can be synthesized. In fact, using only a few of the metal ions from Table 5.2 and several of the ligands in different combinations from Table 5.1, it is apparent that thousands of complexes could arise. Of course, many of the resulting complexes would not yet have been synthesized. The synthesis and characterization of new coordination compounds currently represents an area of intense research interest.

The composition and structures of several common complexes are shown in Figure 5.1. The geometry around a given metal ion is determined principally by the metal ion, but the nature of the ligands can also be important. The charge on each complex ion is determined by summing the charge of the metal ion and of the ligands present.

Composition of the Complex	Net Charge
$Ag^+ + 2NH_3 \rightarrow Ag(NH_3)_2^+$	$(+1) + 2(0) = +1$
$Pt^{2+} + 4Cl^- \rightarrow PtCl_4^{2-}$	$(+2) + 4(-1) = -2$
$Cr^{3+} + 3Cl^- + 3NH_3 \rightarrow Cr(NH_3)_3Cl_3$	$(+3) + 3(-1) + 3(0) = 0$

From the examples above it should be evident that the charge on a complex ion need not be memorized, but can readily be determined for each complex. In the case of central ions with variable oxidation numbers, the oxidation number of the metal ion is required, also. For example, the charge of the hexacyanoferrate(II) ion must be $(+2) + 6(-1) = -4$.

Two of the complexes in Figure 5.1 merit attention at this point. The ammine complex of copper(II) is frequently represented as $Cu(NH_3)_4^{2+}$, yet experimental evidence suggests that two H_2O molecules are also present in the first coordination sphere, that is, $Cu(NH_3)_4(H_2O)_2^{2+}$. The bonds to the four NH_3 molecules in the square plane are relatively short. The bonds to the two H_2O molecules, which occupy *axial* positions above and below the plane, are somewhat longer and are indicated with dotted lines in Figure 5.1.

As noted earlier, complexes of nickel(II) are both six-coordinate and four-coordinate. A four-coordinate, square-planar complex, *bis*dimethylglyoximatonickel(II), is shown in Figure 5.1. Each of the two bidentate anion ligands is derived from the parent molecule, dimethylglyoxime, $HC_4H_7N_2O_2$. The four nitrogen atoms are the donor atoms in the square plane that includes the Ni^{2+} ion. The $Ni(C_4H_7N_2O_2)_2$

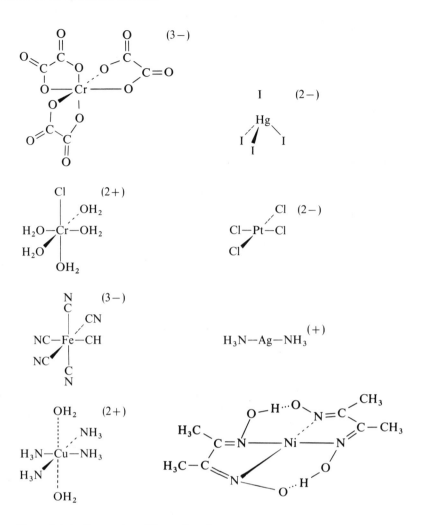

Figure 5.1 | *Geometries of Some Complexes*

complex is neutral and has a very low solubility in water. This, plus the fact that the complex is highly colored, makes dimethylglyoxime (DMGH for short) a useful reagent to test for the presence of nickel(II).

5.4 | ISOMERISM

Two or more coordination compounds that have the same molecular formula but different structural arrangements of atoms or groups are called **isomers**. Many types of isomerism are possible for coordination compounds, four of which are discussed in this section.

Hydrate Isomerism

Compounds in which the number of anions or molecules present exceeds the coordination number of the central ion possess isomers in which the ligands in the first coordination sphere have been interchanged with the remaining anions or molecules. For example, commercially available hydrated chromic chloride has the composition $CrCl_3 \cdot 6H_2O$.

The coordination number of Cr^{3+} is six, but there are a total of nine possible ligands. Two isomers of this compound are

$[Cr(H_2O)_6]Cl_3$ $[Cr(H_2O)_4Cl_2]Cl \cdot 2H_2O$

hexaaquochromium(III) chloride   dichlorotetraaquochromium(III) chloride dihydrate

violet   *green*

The Cl^- ions can either be a part of the first coordination sphere or are simply present in the solid lattice. The process of exchanging ligands bonded to Cr^{3+} is a very slow one in water solution so that the two forms can exist independently. They can be distinguished experimentally on the basis of their colors or by precipitation reactions with Ag^+:

$$[Cr(H_2O)_6]Cl_3 + 3Ag^+ \rightarrow Cr(H_2O)_6^{3+} + 3AgCl(s) \quad (1)$$

$$[Cr(H_2O)_4Cl_2]Cl \cdot 2H_2O + Ag^+ \rightarrow Cr(H_2O)_4Cl_2^+ + AgCl(s) \quad (2)$$

All three chlorides in hexaaquochromium(III) chloride precipitate immediately, whereas only the single, outer chloride ion is precipitated in $[Cr(H_2O)_4Cl_2]Cl \cdot 2H_2O$.

Geometrical Isomerism

Geometric isomers differ in the relative positioning of ligands in a square planar or octahedral complex. At least two different types of ligands must be present. The two geometric isomers of dichlorodiammine-platinum(II) are

cis-Pt(NH$_3$)$_2$Cl$_2$ *trans*-Pt(NH$_3$)$_2$Cl$_2$

The prefix *cis* indicates that identical ligands are placed in adjacent coordination sites, *trans* in opposite sites across the plane. A number of geometric isomers of similar compounds have been prepared and the structures determined by x-ray crystallography. The cis and trans forms usually have somewhat different chemical and physical properties arising from the different geometries. The two NH_3 molecules in *cis*-Pt(NH$_3$)$_2$Cl$_2$ can be replaced by an ethylenediamine molecule to give *cis*-Pt(en)Cl$_2$. In this compound a trans isomer is not possible, because the two nitrogen donor atoms in ethylenediamine must occupy adjacent positions in the square plane.

The octahedral complex ion, $Cr(H_2O)_4Cl_2^+$, described in the previous section, can also exist in cis and trans forms:

cis-Cr(H$_2$O)$_4$Cl$_2^+$ *trans*-Cr(H$_2$O)$_4$Cl$_2^+$

Again, note that the two Cl⁻ ions are at 180° from each other in the trans form, but are in the adjacent, 90° positions in the cis form. Since all six positions in an octahedral complex are identical, the Cl⁻ ions must be located either 90° or 180° from one another.

Optical Isomerism

Complexes that lack any element of symmetry because of the arrangement of the ligands can exist in two isomeric forms. The two forms are called **optical isomers** and differ only in the way in which they interact with polarized light. The $Cr(C_2O_4)_3^{3-}$ ion in Figure 5.1 can exist in two optically active forms, one of which is just the "mirror image" of the other. The two forms cannot be superimposed on one another and, therefore, are related to each other in the same manner as are a right-handed and left-handed glove.

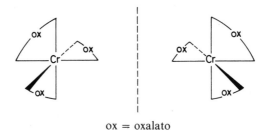

ox = oxalato

Linkage Isomerism

Linkage isomers arise when a given ligand may be bound to the central ion by either of two different donor atoms. In Table 5.1 the donor atom in the thiocyanate, SCN⁻, ligand is indicated as the nitrogen atom. In principle, the atom at either end of the ion could serve as the donor atom. The N-bonded form has been observed in most of the complexes studied, but the S-bonded form does exist, for example in $Pt(NH_3)_2(SCN)_2$. Recently, evidence for both the N- and S-bonded forms of thiocyanatopentaaquochromium(III) ion has been obtained.

Suggested Reading

Basolo, F., and R. C. Johnson, *Coordination Chemistry*, Benjamin, New York, 1964 (paperback).

Jones, M., *Elementary Coordination Chemistry*, Prentice-Hall, Englewood Cliffs, N.J., 1964.

Martin, D. F., and B. B. Martin, *Coordination Compounds*, McGraw-Hill, New York, 1964 (paperback).

Murmann, R. K., *Inorganic Complex Compounds*, Van Nostrand Reinhold, New York, 1964 (paperback).

Problems

1. What ions are present in an aqueous solution of $Na_3[Co(NO_2)_6]$?

2. Write formulas for each complex ion or compound. Place brackets around the complex ion in each coordination compound.

a. Tetraaquozinc(II) ion_____

b. Diamminesilver(I) nitrate_____

c. Ammonium hexafluorocobaltate(III)_____

d. Hexaamminenickel(II) chloride_____

e. Pentacyanoiodocobaltate(III) ion_____

f. Dithiomercurate(II) ion_____

g. Dichloro*bis*ethylenediaminechromium(III) ion_____
(The prefix *bis*, which indicates 2, is used in preference to *di* for chelating ligands and for ligands with names that begin with a Greek prefix, for example dimethylglyoxime.)

3. Write systematic names for each complex ion or coordination compound.

a. $Fe(H_2O)_6^{2+}$_____

b. $K_2[PtBr_4]$_____

c. $Na_2[CdCl_4]$_____

d. $Cr(NH_3)_3(SCN)_3$_____

e. $Co(NH_3)_5SO_4^+$_____

f. $Pt(en)Cl_2$_____

g. $K[Cr(H_2O)_2(C_2O_4)_2]$_____

4. Indicate the correct charge on the complex ion, based on the charges of the component ions and molecules.

a. $Ag(S_2O_3)_2$ ☐   *b.* $Cr(H_2O)_2(C_2O_4)_2$ ☐   *c.* $Ni(en)_3$ ☐

5. What is the oxidation number of the central metal ion?

a. HgI_4^{2-}_____   *b.* FeF_6^{3-}_____

c. $Pt(NH_3)_2(SCN)_4$_____

6. Write formulas for the hydrate isomers of $Ni(H_2O)_4Cl_2 \cdot 2H_2O$.

7. Sketch structures for each of the geometric isomers of the following coordination compounds. The sketches should clearly indicate the three-dimensional arrangement of ligands around the central metal ion.

a. $Pt(H_2O)_2Cl_2$

Coordination Compounds of Transition Metals

    b. $PtBrCl_2I$

    c. $Co(NH_3)_4Cl_2$

    d. $Cr(H_2O)_3F_3$

8. Demonstrate that the tetrahedral $HgBr_2I_2^{2-}$ ion does not possess any geometric isomers.

9. Make a sketch of the linkage isomers involving the NO_2^- ion in $Co(CN)_5NO_2^{3-}$. (When NO_2^- is bonded to the metal ion through N, the name *nitro* is used; when NO_2^- is bonded through one of the oxygen atoms, the name *nitrito* is used.)

Properties of Complex Ions

CHAPTER 6

6.1 | BONDING

Bonds between ligands and metal ions are comparable in energy with normal ionic and covalent bonds. The nature of the metal–ligand bond depends upon the electronic structures of the ligands and of the transition metal cation involved. In particular, the partially filled d orbitals of transition metal ions play an important role in determining the characteristics of complex ions.

Electron Configurations

The d-electron configurations of the first-row transition metal ions are shown in Table 6.1. These ions arise from elements with atomic numbers 21 through 30 in the periodic table. Ions that are not stable in water solution, such as Sc^{2+}, Mn^{3+}, and Ni^{3+}, are omitted. The electron configurations in Table 6.1 are based on the electron configurations of the atoms and assuming loss of $4s$ electrons in the process of ion formation. For example, the electron configuration of an isolated atom of iron, having an atomic number of 26, is $[Ar]4s^2 3d^6$. (The notation $[Ar]$ represents the completed, 18-electron structure of the element argon.) Loss of the two $4s$ electrons results in the $Fe^{2+}(3d^6)$ ion and further loss of a $3d$ electron results in the $Fe^{3+}(3d^5)$ ion.

Ions in the $+2$ oxidation state occur in aqueous solution for all the first-row transition elements except scandium. The $+3$ state also occurs for several of the elements. Although the $+2$ ions are anticipated on the basis of loss of two $4s$ electrons, it is not possible in

Table 6.1 | *Valence-Electron Configuration of Ions of 3d Elements*

									Cu$^+$	
M$^+$									$3d^{10}$	
M^{2+}		Ti^{2+}	V^{2+}	Cr^{2+}	Mn^{2+}	Fe^{2+}	Co^{2+}	Ni^{2+}	Cu^{2+}	Zn^{2+}
		$3d^2$	$3d^3$	$3d^4$	$3d^5$	$3d^6$	$3d^7$	$3d^8$	$3d^9$	$3d^{10}$
M^{3+}	Sc^{3+}	Ti^{3+}	V^{3+}	Cr^{3+}		Fe^{3+}	Co^{3+}			
	$3d^0$	$3d^1$	$3d^2$	$3d^3$		$3d^5$	$3d^6$			

general to predict what other oxidation states will be observed for a given element. However, for elements 21 through 25, the maximum oxidation state observed is equal to the total number of 4s and 3d electrons. These maximum oxidation states are Ti(IV), V(V), Cr(VI), and Mn(VII). The CrO_4^{2-} ion is an example of chromium in the +6 state and MnO_4^- of manganese in the +7 state. Several intermediate states exist; for example, V(IV) and Mn(IV) are stable in water solution, although Cr(IV) is not. A more detailed discussion of the various oxidation states of several of these elements is included in Part II.

Metal Ion–Ligand Bonding

Bonds between ligands and the central metal ion are partly ionic and partly covalent in character. The bonding in a number of complexes arises primarily from electrostatic attractions between the positive central ion and the negative portions of the ligands. A consideration of the electrostatic interactions, together with the effect of the ligands on electrons in d orbitals, forms the basis of **crystal field theory**. In accounting for bond energies and other properties of complexes, the sharing of electrons between ligands and the metal ion must also be considered. Covalent bonding is especially important for ligands with heavier donor atoms, such as phosphorus and sulfur, and for ligands such as CN^- and CO. For metal ions of the first-row transition elements, 4s, 4p, and 3d orbitals are involved in the formation of covalent bonds. **Valence-bond theory** describes bonding in octahedral complexes using d^2sp^3 hybrid orbitals formed from these atomic orbitals. A newer and more useful theory is the **molecular orbital approach**, which involves sharing of electrons in molecular orbitals arising from combinations of the ligand orbitals with the 4s, 4p, and 3d orbitals of the metal ion.

The nature of the bonding in complex ions is similar to that in ordinary molecules. The details of the bonding theories will not be considered here; instead, attention will be focused on the way in which the ligands affect the behavior of 3d electrons in complex ions.* The general aspects of crystal field theory are described below for octahedral complexes. Similar descriptions, although somewhat different in detail, can be applied to complexes with other geometries.

Crystal Field Theory

A model of an octahedral complex can be constructed by placing ligands on the x, y, and z axes of a rectangular coordinate system at the center of which is placed the metal ion. In Figure 6.1 the ligands are represented as open circles with a negative charge. This symbol represents either the unit negative charge on a ligand such as Cl^- or the negative end of a dipolar molecule such as H_2O. A possible set of five 3d orbitals is shown in Figure 6.2, in which the $3d_{z^2}$ and $3d_{x^2-y^2}$ orbitals arbitrarily are assumed to lie along the z and xy axes, respectively. Figure 6.2 shows the space orientation of electrons in the 3d orbitals. In a metal ion isolated in the gas phase, a 3d electron may occupy any

---
* For further information, consult the suggested references at the end of Chapter 5.

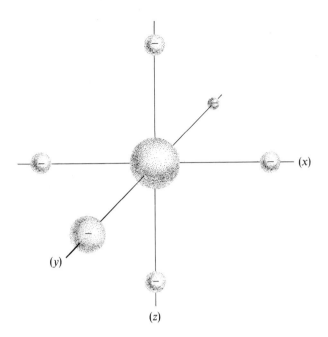

Figure 6.1 | *Model of an Octahedral Complex*

of the 3d orbitals, since all are of the same energy. However, placing six identical ligands around the central metal ion as shown in Figure 6.1 has two important effects on electrons in the 3d orbitals. First, since the donor atoms in the ligands bear at least a partial negative charge, all the 3d electrons are repelled by the ligands. Thus, the energy levels of all the 3d orbitals increase (become less stable) in the complex, relative to the free metal ion. Second, electrons in the $3d_{z^2}$ and $3d_{x^2-y^2}$ orbitals interact more strongly with the ligands, because these orbitals

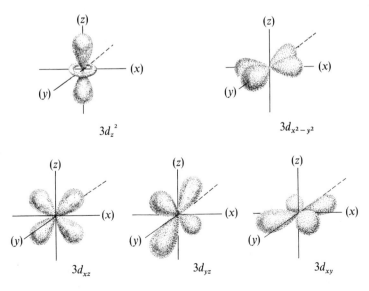

Figure 6.2 | *3d Orbitals*

point directly at the ligands which are located on the x, y, and z axes. Electrons in the remaining $3d_{xz}$, $3d_{xy}$, and $3d_{yz}$ orbitals interact less strongly with the ligands, because these orbitals are oriented between the axes. Thus, the ligands bring about a splitting in the energy levels of the 3d orbitals as shown in Figure 6.3. The $3d_{z^2}$ and $3d_{x^2-y^2}$ orbitals, which make up the higher energy level in Figure 6.3, are referred to as e_g *orbitals*. The remaining orbitals, which make up the lower energy level, are referred to as t_{2g} *orbitals*. The separation in energy between the two levels is usually denoted as 10 Dq. Since there are two sets of levels in an octahedral complex, the electrons prefer to be in the orbitals of lower energy. This situation has important consequences for the properties of these complexes.

6.2 | SPECTRAL AND MAGNETIC PROPERTIES

Colors of Complex Ions

The model developed in Section 6.1 suggests that octahedral complexes should be capable of absorbing energy. An electron in the t_{2g} level can be excited to the e_g level by absorption of energy of the appropriate wavelength. Such absorption is observed, and for many complex ions, the wavelengths involved lie in the visible region of the spectrum from 4000 to 7000 Å. These electronic absorptions are responsible for the color of many transition metal complexes. (Other modes of light absorption exist also.) The simplest example is the $Ti(H_2O)_6^{3+}$ complex ion, in which the Ti^{3+} ion possesses a single 3d electron. This ion absorbs light in the region of 4930 Å and, therefore, is violet in color. For metal ions with more than one 3d electron, repulsion between electrons creates the possibility of several excited states and a correspondingly more complicated spectrum. For example, the violet $Cr(H_2O)_6^{3+}$ ion, which contains three 3d electrons, absorbs energy at 2600, 4100, and 5800 Å.

The magnitude of 10 Dq and, therefore, the energy of the light absorbed depend on the nature of the metal ion and the ligands. For most +2 ions of the first-row transition metals, the sequence of increasing

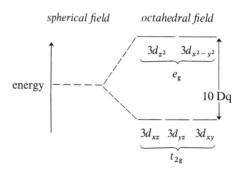

Figure 6.3 | *Splitting of 3d Energy Levels in an Octahedral Ligand Field*

ligand field strength (increasing values of 10 Dq) is

$$I^- < Br^- < Cl^- < H_2O < SCN^- < NH_3 < CN^-$$

Thus, changes in the ligands present in the first coordination sphere can be expected to result in changes of color in complex ions.

Electron Configurations and Magnetic Properties

In assigning electrons to orbitals in the t_{2g} and e_g energy levels, two factors of primary importance must be kept in mind. First, electrons normally occupy the more stable orbitals, which are lower in energy. Second, electrons tend to avoid being paired in a single orbital. Two electrons constrained to occupy the same region in space repel one another more than two electrons in separate orbitals. As a consequence, the electron configurations of complex ions with one, two, or three 3d electrons are as shown in Figure 6.4(a). A complex ion with four 3d electrons can have either a $(t_{2g})^3(e_g)^1$ or a $(t_{2g})^4$ configuration, as shown in Figure 6.4(b). When the ligands present have high ligand field strengths, the **low-spin configuration**, $(t_{2g})^4$, will occur. This situation results because the excitation energy $t_{2g} \to e_g$, caused by the strong-field ligands, is larger than the unfavorable electron-pairing energy. Conversely, weak-field ligands lead to the **high-spin configuration**, because in these complexes the unfavorable pairing energy can be avoided without too great a cost in excitation energy.

The presence of unpaired electrons can be detected experimentally by observing the behavior of a sample of a coordination compound in a strong magnetic field. Compounds containing one or more unpaired electrons are attracted by a magnetic field and are said to be **paramagnetic**. Compounds containing no unpaired electrons are slightly repelled by a magnetic field and are said to be **diamagnetic**. Samples containing complex ions with one, two, or three 3d electrons, as in Figure 6.4(a), will be paramagnetic to the extent of one, two, or three unpaired electrons,

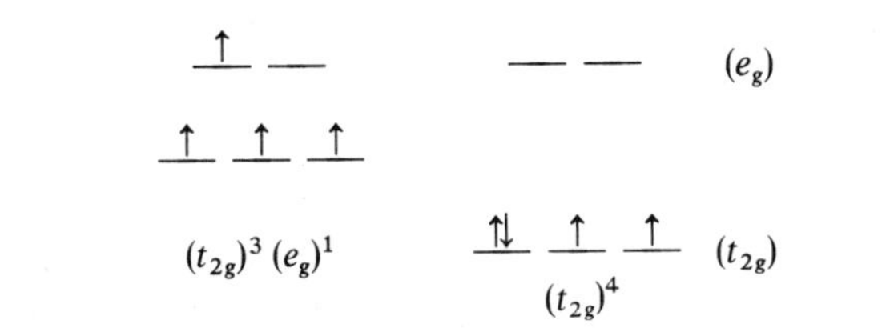

Figure 6.4 | *3d-Electron Configurations in Complex Ions*

respectively. An ion with a $(t_{2g})^3(e_g)^1$ configuration will be paramagnetic to the extent of four electrons, while an ion with a $(t_{2g})^4$ configuration will be paramagnetic to the extent of only two unpaired electrons. Thus, magnetic measurements provide a means of experimental verification of some of the predictions derived from crystal field theory.

6.3 REACTIONS OF COMPLEX IONS

Ligand Dissociation

Complex ions have a tendency to dissociate in water solution. The dissociation of the diammine silver(I) ion is an example of this type of reaction:

$$Ag(NH_3)_2^+ \rightleftarrows Ag^+ + 2NH_3 \qquad (1)$$

The corresponding equilibrium constant

$$K = 6 \times 10^{-8} = \frac{[Ag^+][NH_3]^2}{[Ag(NH_3)_2^+]} \qquad (2)$$

is called an **instability constant**. Its value provides a means of establishing conditions in which $Ag(NH_3)_2^+$ will be the predominant silver species in solution. For example, when excess NH_3 is present in $1\,M$ concentration, the ratio $[Ag^+]/[Ag(NH_3)_2^+]$ has a value of 6×10^{-8}. Thus, in $1\,M$ NH_3, silver ion is essentially completely converted to the diammine complex ion.

Equation (1) describes the overall dissociation of $Ag(NH_3)_2^+$, but the actual process involves loss of, first, one NH_3 to give $Ag(NH_3)^+$, followed by loss of the second NH_3 to give Ag^+. Ligand dissociation is really a substitution process in which water molecules replace the ligands present:

$$Ag(NH_3)_2^+ + 4H_2O \rightleftarrows Ag(H_2O)_4^+ + 2NH_3 \qquad (3)$$

(Experimental evidence suggests that the hydrated silver ion is surrounded by four, rather than two, H_2O molecules.) Calculations based on equation (2) may be made without taking specific account of the water molecules, since the concentration of water is included in the instability constant.

Ligand Substitution Reactions

Replacement of one ligand in the first coordination sphere by another ligand is called a **substitution reaction**. Several examples of substitution reactions are shown in Figure 6.5. Substitution leaves the oxidation state of the metal ion unchanged. However, changes in bonding occur and, accompanying the introduction of a new ligand with a different ligand field strength, changes in color frequently result as well. An important area of current research involves characterization of new substitution reactions. In addition to the stoichiometries of the reactions, reaction rates and the factors that influence the rates are being studied in detail. The nature of a substitution reaction depends

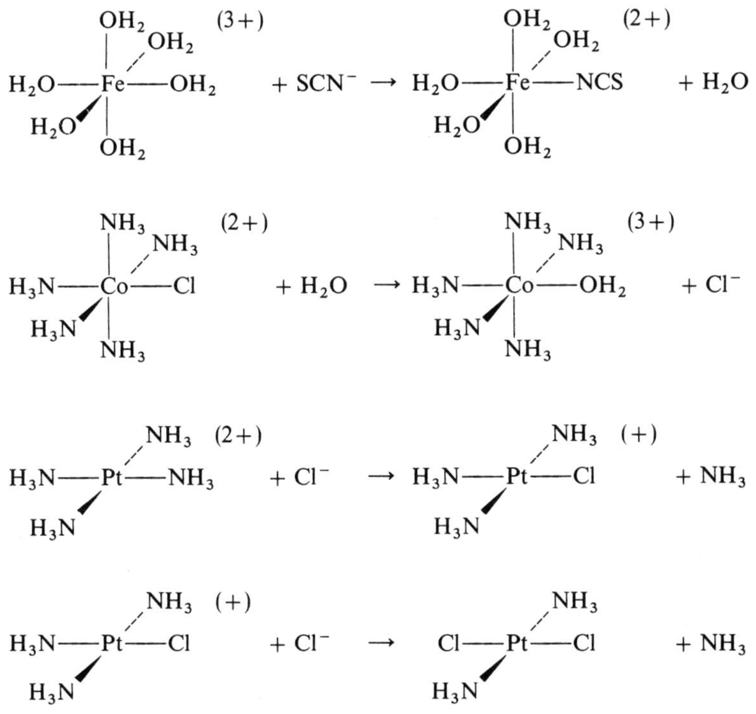

$$\underset{\substack{H_2O \\ H_2O}}{\overset{\substack{OH_2 \\ OH_2}}{Fe}}-OH_2 \quad (3+) + SCN^- \rightarrow \underset{\substack{H_2O \\ OH_2}}{\overset{\substack{OH_2 \\ OH_2}}{Fe}}-NCS \quad (2+) + H_2O$$

$$\underset{\substack{H_3N \\ NH_3}}{\overset{\substack{NH_3 \\ NH_3}}{Co}}-Cl \quad (2+) + H_2O \rightarrow \underset{\substack{H_3N \\ NH_3}}{\overset{\substack{NH_3 \\ NH_3}}{Co}}-OH_2 \quad (3+) + Cl^-$$

$$\underset{H_3N}{\overset{NH_3}{Pt}}-NH_3 \quad (2+) + Cl^- \rightarrow \underset{H_3N}{\overset{NH_3}{Pt}}-Cl \quad (+) + NH_3$$

$$\underset{H_3N}{\overset{NH_3}{Pt}}-Cl \quad (+) + Cl^- \rightarrow \underset{H_3N}{\overset{NH_3}{Cl-Pt}}-Cl \quad + NH_3$$

Figure 6.5 | *Ligand Substitution Reactions*

on the metal ion, the structure of the complex, and the type of ligands present. An interesting example of this point is provided by the replacement of NH_3 molecules by Cl^- ions in the square-planar platinum(II) complexes shown in Figure 6.5. Note that the reaction of $Pt(NH_3)_3Cl^+$ with Cl^- produces the trans, rather than the cis, complex of $Pt(NH_3)_2Cl_2$. Apparently, the nature of the existing ligands determines the structural course of the reaction.

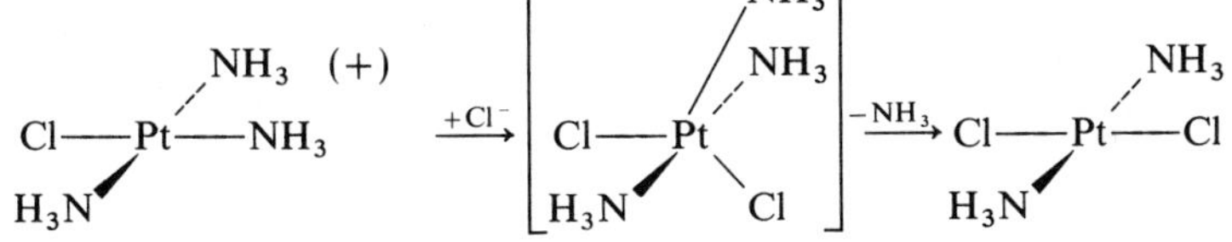

$$\underset{H_3N}{\overset{NH_3}{Cl-Pt}}-NH_3 \quad (+) \xrightarrow{+Cl^-} \left[\underset{H_3N}{\overset{NH_3, NH_3}{Cl-Pt}}-Cl\right] \xrightarrow{-NH_3} \underset{H_3N}{\overset{NH_3}{Cl-Pt}}-Cl$$

(a)

$$Co(CN)_5H_2O^{2-} \rightleftarrows Co(CN)_5^{2-} + H_2O$$
reactant   intermediate

$$Co(CN)_5^{2-} + I^- \rightleftarrows Co(CN)_5I^{3-}$$
product

(b)

Figure 6.6 | *Possible Mechanisms for Substitution:*
(a) Platinum(II) Complex; (b) Cobalt(III) Complex

It is interesting to inquire about the ways in which a substitution process can occur. One **mechanism** that can be envisioned is the simultaneous breaking of the bond to the old ligand as the bond to the new ligand is being formed. Square-planar platinum(II) complexes are assumed to react in this fashion, as shown in Figure 6.6(a). Substitution in a platinum(II) complex is referred to as a **bimolecular process** because the progress of the reaction depends on the concentrations of both the complex ion and the entering ligand. Another possible mechanism for substitution involves essentially complete breaking of the bond to the old ligand before the bond to the new ligand is formed. An example of a reaction thought to occur by this type of **unimolecular process** is shown in Figure 6.6(b). However, substitution reactions of octahedral complexes are currently being studied because the mechanisms usually are not as simple as that shown in Figure 6.6(b). Apparently, some bond formation to the new ligand occurs before the bond to the old ligand is completely broken.

Oxidation–Reduction Reactions

A number of reactions of complex ions involve a change in the oxidation state of the central metal ion. Certain of these reactions occur without a change in the ligands present and are called **electron-transfer reactions**. An example of this type of reaction is

$$2Fe(CN)_6^{4-} + I_2 \rightarrow 2Fe(CN)_6^{3-} + 2I^- \tag{4}$$

In equation (4) a change from iron(II) to iron(III) takes place. This process may be thought of as involving the transfer of two electrons from two $Fe(CN)_6^{4-}$ ions to the I_2 molecule.

Oxidation–reduction reactions of complex ions often occur with a change in both the number and type of ligands in the first coordination sphere. The oxidation of chromium(III) to chromium(VI) by hydrogen peroxide in basic solution is an example of this type of reaction:

$$2Cr(H_2O)_2(OH)_4^- + 3H_2O_2 + 2OH^- \rightarrow 2CrO_4^{2-} + 12H_2O \tag{5}$$

Reaction (5) is a complex process involving a series of ligand replacements as well as changes in oxidation states.

Problems

1. Write out the electron configuration of the element and of the corresponding ions indicated.

   a. Cr_____
      Cr²⁺_____
      Cr³⁺_____

   b. Cu_____
      Cu²⁺_____

   c. Ag_____
      Ag⁺_____

   d. Ru_____
      Ru³⁺_____

2. The four ligands in a tetrahedral complex ion can be assumed to occupy positions in the octants in between the x, y, and z axes rather than on these axes. Draw an energy-level diagram, similar to that in Figure 6.3, which shows the relative splitting of the $3d$ energy levels in a tetrahedral ligand field.

3. If Br^- ions were substituted for H_2O molecules in the first coordination sphere, would the wavelength of the absorption maximum of the $Ti(H_2O)_6^{3+}$ ion be expected to increase or to decrease? Explain. (Wavelength and energy are inversely proportional to each other.)

4. a. Show the high-spin and the low-spin $3d$ electron configuration in a d^6 complex, using the form in Figure 6.4.

   b. Which compound, $K_4[Fe(CN)_6]$ or $[Fe(H_2O)_6]SO_4 \cdot H_2O$, is expected to be paramagnetic and how many unpaired electrons are present in the complex?

Properties of Complex Ions

5. For the ligand dissociation $Ag(NH_3)_2^+ \rightleftarrows Ag^+ + 2NH_3$, $K = 6 \times 10^{-8}$. In a solution containing excess $0.1\ M$ NH_3, calculate the value of the concentration ratio $[Ag^+]/[Ag(NH_3)_2^+]$.

6. Write equations that show the stepwise dissociation of $Ag(S_2O_3)_2^{3-}$.

7. For the dissociation of the iron(III)–thiocyanate complex, $Fe(H_2O)_5NCS^{2+} + H_2O \rightleftarrows Fe(H_2O)_6^{3+} + SCN^-$, $K = 7 \times 10^{-3}$. In a solution containing excess $0.5\ M$ SCN^-, calculate the value of the ratio $[Fe(H_2O)_6^{3+}]/[Fe(H_2O)_5NCS^{2+}]$.

8. Write structural equations like those shown in Figure 6.5 for the two reactions involved in the conversion of $PtCl_4^{2-}$ to cis-$Pt(NH_3)_2Cl_2$ by the addition of NH_3.

PART II | Cation Chemistry and Procedures

Guide to Experimental Work

CHAPTER 7

7.1 THE FIRST LABORATORY PERIOD

The overall plan of experiments to be performed is outlined in Chapter 1. To aid in making an efficient start on these experiments, specific directions for the first laboratory period are listed below.

1. *Equipment.* If this is your first period in laboratory, check the equipment assigned to you. Use the check list of equipment provided by the instructor. Report to the instructor any missing or damaged items. They will be replaced without charge during the first period. (A recommended list of equipment and supply items is shown on page 188.)
2. *Supplies.* Obtain a polyethylene wash bottle from the stockroom (optional). Also, from the stockroom, obtain a package of supplies for the cation–anion experiments. This package contains expendable items not found in the lab desk.
3. *Stirring rods.* If thin (3 mm) glass stirring rods are not available in your desk, they should be made from solid rod obtained in the stockroom. Cut the rods into 4- to 5-inch lengths. Use a file to scratch the glass rod where it is to be cut. Place your thumbs opposite the scratch and snap the glass with an outward pressure. Fire polish both ends of each rod so that no sharp edges remain. Heat the end of the rod by rotating it in a hot Bunsen burner flame. After a rounded surface has formed, allow the glass rod to cool on a piece of asbestos.
4. *Droppers.* Glass droppers with rubber bulbs are used for the addition of reagent solutions. These droppers should deliver approximately 20 drops per milliliter. Calibrate several of the droppers using a 10-ml graduated cylinder. Fill the cylinder exactly to the 1.0-ml level with distilled water. Then, using a dropper, count the number of drops of water required to raise the level to the 2.0-ml mark. If the droppers you have deliver less than 15 or more than 25 drops per milliliter, a correction factor should be applied in the experimental procedures. For example, using a dropper giving 30 drops per milliliter, the directions to add 4 drops of a reagent should be adjusted to 6 drops of reagent.

(Optional.) If glass droppers are not available, they can be constructed as follows. Cut a piece of 6-mm or 7-mm soft glass tubing into 6-inch lengths. Rotate the center of the piece of tubing in the hot tip of a Bunsen burner flame until the glass is red-hot and begins to sag. Remove the tube from the flame, pull the ends apart slowly, and then more rapidly as the tubing cools. When the glass has cooled, remove the thin center section by cutting the glass in the two tapered regions. The large end of each of the dropper tubes should be heated until the glass is soft. Use the tip of a file to flare out the end. Reheat the flared end and then press it quickly on an asbestos pad to form a flange. When the glass is cool, place a rubber bulb on the end of the dropper tube. Briefly heat the delivery tip to remove the sharp edges. A total of four droppers should be sufficient. It is recommended that one of the droppers be cut so that it has a long, narrow delivery tip. This dropper will be useful in withdrawing a small volume of liquid from a precipitate in a 3-inch test tube.

5. *Desk set of reagents.* Your desk may contain a set of small bottles fitted with glass droppers. These bottles are used to dispense frequently used reagent solutions, referred to here as **primary solutions**. If the individual desk sets are not available, all necessary solutions will be found in the reagent area. (In this case, omit the steps below.)

   Clean the small reagent bottles. Discard any old solutions found in the bottles. Rinse the bottles thoroughly with distilled water. Replace any of the bottles, or droppers, which cannot be cleaned satisfactorily. Unless the labels are of a permanent type, remove old labels with hot water. Relabel the bottles with the concentration and formula of reagent solutions in the **primary reagent** list on page 73. From supply bottles in the reagent area, fill each bottle one-half to two-thirds full. (CAUTION: See safety rules!) Keep the dropper tops screwed in place except during use.

6. *Preliminary information.* Before starting experimental work, read the important sections below: Laboratory Safety, Laboratory Reagents, and Experimental Techniques.

7. *First experiments.* Begin the experimental procedures for Group 1 in Chapter 8.

7.2 LABORATORY SAFETY

The experiments involved are safe if several basic safety precautions are taken. Please, for your own safety, learn the rules below and practice them at all times. Consult the instructor concerning safety and first-aid measures in effect in the laboratory. Report any accident or injury to the instructor.

1. *Corrosive reagents.* Acids and bases cause skin damage. Bases such as $NaOH$ and NH_3 are especially hazardous because they act so rapidly. Contact with acid and base solutions must be avoided.

   Keep your face away from anything that contains an acid or base.

Immediately wash off with water any reagents spilled on the skin. If an acid or base solution should splash into your eye, get *immediate* help in washing it out.

Do not pipet any liquid by mouth. Use a pipet bulb. Most metal ion solutions are toxic.

2. *Safety glasses.* Wear safety glasses or some form of eye protection at all times in the laboratory. This is imperative when you are heating mixtures containing acids or bases.
3. *Heating test tubes.* Never, under any circumstances, heat a test tube directly in a flame. Always place it in a hot-water bath, even though the procedure may simply specify that a given solution be heated. Keep the open end of the test tube pointed away from you and your neighbors. Wear safety glasses when mixtures with acids and bases are being heated.

7.3 REAGENTS

All the reagents required for the experiments are kept in the reagent supply area. The more frequently used reagents are designated primary solutions. These should be used to fill your desk set of dropper bottles, if these bottles are available. It is important that all reagent bottles remain free of contamination. Please help maintain the integrity and availability of the reagents by observing the following rules:

1. Leave laboratory reagent bottles on the shelves and in proper order. Do not take them to your desk.
2. Put dropper tops back onto the correct bottle immediately after use. Notify the instructor if the dropper tops get mixed.
3. Do not pour unused reagent back into the supply bottle.
4. Do not let the reagents get up into the rubber bulbs in the dropper bottles. Dropper bottles in the desk set should not be allowed to tip over.

Primary Solutions

$2\,M$ HCl $6\,M$ HCl $3\,M$ NH_4Cl $2\,M$ NH_3 $6\,M$ NH_3 $1\,M$ HNO_3
$6\,M$ HNO_3 $6\,M$ $HC_2H_3O_2$ Thioacetamide $0.1\,M$ K_2CrO_4

Secondary Reagents

$0.1\,M$ $AgNO_3$	$12\,M$ HCl
$0.03\,M$ $Hg_2(NO_3)_2$	$18\,M$ H_2SO_4
$0.1\,M$ $Pb(NO_3)_2$	$6\,M$ NaOH
$0.05\,M$ As(III)	$2\,M$ $NH_4C_2H_3O_2$
$0.1\,M$ $Hg(NO_3)_2$	$2\,M$ $(NH_4)_2CO_3$
$0.1\,M$ $Cu(NO_3)_2$	$0.5\,M$ $(NH_4)_2C_2O_4$
$0.1\,M$ $Cd(NO_3)_2$	$0.1\,M$ NH_4SCN
$0.1\,M$ $Fe(NO_3)_3$	$0.1\,M$ $Pb(C_2H_3O_2)_2$
$0.1\,M$ $Ni(NO_3)_2$	$0.05\,M$ $K_4Fe(CN)_6$
$0.2\,M$ $Al(NO_3)_3$	$0.2\,M$ $SnCl_2$
$0.1\,M$ $Cr(NO_3)_3$	Aluminon
$0.1\,M$ $Ba(NO_3)_2$	3 percent H_2O_2
$0.2\,M$ $Ca(NO_3)_2$	Dimethylglyoxime
$0.3\,M$ NH_4NO_3	Sodium dithionite(s)
$0.1\,M$ $NaNO_3$	$MnO_2(s)$

7.4 | EXPERIMENTAL TECHNIQUES

Before beginning experimental work, read through the list of techniques in this section. Proper handling of reagents and reaction mixtures will save time and avoid errors in identification. Review this section occasionally.

1. *Working area.* Spread out a towel on part of the desk top. Clean equipment should be placed on this towel, ready for use. Keep the remainder of the working area clean. Use a wet sponge to wipe up solutions spilled on the desk top.
2. *Cleaning equipment.* Glass droppers and stirring rods should be rinsed *immediately* after use with distilled water. Pipe cleaners are useful for removing solid particles clinging to the insides of glass dropper tubes. After use, test tubes should be rinsed once or twice with tap water and then filled with water. Do not let solid particles on the glass walls dry out and harden. When several test tubes have accumulated, clean them thoroughly with a small test tube brush. Rinse the tubes with distilled water.
3. *Distilled water.* Use only distilled water to add to reaction mixtures or to wash precipitates. The use of a medium-sized beaker and a pipet dropper, or of a polyethylene wash bottle, is recommended for dispensing distilled water. **Hot-water bath.** At the beginning of the laboratory period, fill a small beaker (100 or 150 ml) half full of distilled water. Heat the water and maintain its temperature near the boiling point with a low burner flame. Do not use tap water, which leaves a residue as it boils away. The hot-water bath can be used to heat test tubes. The tubes can be leaned against the walls of the beaker if the water level is kept at 1 to $1\frac{1}{2}$ inches. If provided in your desk equipment, use the test tube holder plate to keep tubes in an upright position. This metal plate has several holes punched in it to accommodate small test tubes.
4. *Handling reactant solutions.* The experimental procedures involve small volumes of solution, usually about 0.5 ml. Small, 3-inch test tubes should be used to avoid loss of samples. Clearly mark each test tube to avoid confusing one solution with another. Test tubes with white frosted-glass areas can be marked with pencil. Small labels or grease pencils are useful for plain test tubes. Solutions to be stored for several days should be tightly corked. Do not heat a solution in a test tube over an open flame; use a water bath.
5. *Addition of reagents.* Add the specified volume of reagent; do not add an excess. Stir thoroughly after addition, because solutions in small test tubes form layers, rather than mixing spontaneously. Do not allow the reagent dropper tip to come into contact with the sample or the sidewalls of the test tube. Keep your face above and away from reagent droppers as you use them.
6. *Adding acids and bases. pH test paper.* Acid or base solutions should be added one drop at a time. After the addition of each drop, stir thoroughly with a vertical mixing motion. Check the pH with test paper. Do not dip the paper into the solution. Tear each strip of

paper into several small sections. Get a drop of the sample on the end of a stirring rod and place it on the paper. It is important that any reagent adhering to the walls of the test tube be mixed down into the solution prior to making the test.

Litmus paper is satisfactory for distinguishing between acidic or basic solutions. Litmus is red in acid solution and blue in basic solution. Wide-range pH test paper is preferable to litmus paper. It gives an indication of the pH range of the sample as shown below. Use this wide-range paper if it is available.

red	*orange-yellow*	*green*	*blue-green*
strongly acid	*weakly acid*	*weakly basic*	*strongly basic*

7. ***Use of the centrifuge.*** Solids can be separated from solutions conveniently by the use of a centrifuge. A small laboratory centrifuge provides a settling force 200 to 300 times that of gravity. About 10 seconds of high-speed centrifugation will completely settle most precipitates. When using the centrifuge, balance each sample tube with another test tube containing the same volume of liquid. It is usually possible to slow down the centrifuge either with a brake provided or with your hands placed on the smooth, revolving head. Such braking should be done evenly to avoid dispersing the precipitate with a sudden jolt.

    Ideally, the result of centrifuging should be a tightly packed precipitate on the bottom and a clear solution on top. (Clear does not mean the same thing as colorless; it means that no solid remains suspended.) Particles, or a solid film adhering to the sidewalls, are usually not a problem because they remain in the test tube with the precipitate. Particles on the top surface can be removed with a glass dropper or by using a stirring rod to break the curved surface of the liquid. The liquid layer can be separated by pouring it into another test tube. The last drop or two may cling to the lip of the test tube and can be removed by sharp taps with your finger. A narrow-tip glass dropper is useful in removing small amounts of liquid from the precipitate.

8. ***Washing precipitates.*** After pouring off the liquid layer, the remaining precipitate contains a considerable amount of solution absorbed on its surface. A good separation requires that this liquid be removed. The precipitate should be washed with water or the specified wash solution. Add the wash solution to the precipitate and stir thoroughly. Centrifuge and pour off the wash solution. Two or three separate washings are much more effective than a single washing with a large volume. Do not skip the washing steps. Failure to wash completely can waste time by causing confusion in later steps.

9. ***Discarding solutions.*** Do not throw away solutions, or precipitates, unless you are certain that they are not required in a future step. This is especially important in the general unknown. Wash solutions can be discarded unless the directions specify otherwise.

10. ***Heating solutions in an evaporating dish.*** The volume of a solution can be reduced by heating the solution in an evaporating dish.

Gaseous substances, such as NH_3 or H_2S, can be removed in this way also. Hold the dish with a pair of tongs and pass the dish back and forth over the flame. Keep your face away from the dish.

7.5 } USING THIS LABORATORY TEXTBOOK

The following guidelines will help you to make effective use of this book.

1. *Group procedures.* Before coming to each laboratory period, read through the experimental procedures you plan to cover. This preliminary reading will save time in the laboratory and aid in understanding the chemistry involved. The instructor may set up a schedule for the completion of unknowns. You might fall behind at first, but as you become familiar with the techniques, the work will progress more rapidly. Do not rush through the procedures. Take the time you need to understand what is happening in each step.

2. *Group diagrams.* The purpose of each group diagram is to show how the procedural steps are related to each other. As you work through the steps you should know exactly where you are in the group diagram. It is helpful to think in terms of the diagrams whenever possible.

    Reagents to be added are indicated in brackets. In each step the resulting precipitate is shown on the left, the resulting solution on the right. The various precipitates are designated in the diagram, and correspondingly in the procedures, as **P1**, **P2**, etc., the solutions as **S1**, **S2**, etc. Each precipitate or complex ion that confirms the presence of an ion is enclosed in a rectangle.

3. *Recording observations.* Record the results of each major step immediately after that step has been carried out. Use the space provided for Experimental Results adjacent to the group procedure for this purpose. The notes you make can be brief and in any style you like. However, be sure that they are specific and cover the important points. Record the color, if any, of each solution. Indicate the general nature and color of each precipitate. For example, decide which of the following are appropriate: coarse or finely divided, dense and compact or light and fluffy, powdery or thick and gelatinous. Be specific about the color. For example, "yellow" precipitates may range in color from gold to off-white.

    It is recommended that the notations **P1**, **P2**, **S1**, etc., appear in the Experimental Results column and that each of the corresponding precipitates or solutions be described. Make the information recorded as helpful to you as possible. The results should be useful in answering questions at the end of each chapter and exam questions. The group knowns are intended for you to see how each reaction takes place and what the products look like. Whenever a question arises in working on an unknown, refer back to the results of the known. It should be possible to answer many of these questions by a careful comparison of the results for a known ion and the behavior of the unknown.

4. *Group chemistry.* A brief description of the aqueous solution chemistry of the ions in each group is presented in Chapters 8 through 11. The areas covered relate directly to the experimental work you are doing in the laboratory. Read these sections along with the corresponding procedures. These sections should help relate the general principles in Part I to the specific chemistry of the individual ions. It might also be useful to read the corresponding sections in your general chemistry textbook.

Group 1: Ag^+, Pb^{2+}, and Hg_2^{2+}

CHAPTER 8

8.1 CHEMISTRY OF GROUP 1

Group Separation

Silver, lead, and mercury(I) ions share the characteristic of forming insoluble chloride salts. This characteristic, which is unique among the common cations, makes possible the separation of Ag^+, Pb^{2+}, and Hg_2^{2+} from a mixture of many cations. These cations react with a source of chloride ions, such as HCl, to produce the insoluble metal chlorides:

$$Ag^+ + Cl^- \rightarrow AgCl(s) \qquad (1)$$

$$Pb^{2+} + 2Cl^- \rightarrow PbCl_2(s) \qquad (2)$$

$$Hg_2^{2+} + 2Cl^- \rightarrow Hg_2Cl_2(s) \qquad (3)$$

In a mixture of Group 1 ions, these reactions occur together, provided sufficient Cl^- is available. Equations (1), (2), and (3) are important because they express, in a concise way, the chemical results of addition of HCl to Group 1 ions. Before individual ions can be identified, it is necessary to separate the ions. A knowledge of the chemistry of the individual ions is essential for this purpose.

Silver

Only Ag^+, silver in the $+1$ oxidation state, is important in aqueous solution chemistry. However, Ag^+ is a reasonably good oxidizing agent, comparable with I_2 and Fe^{3+}. Thus, in the presence of reducing agents, Ag^+ may be reduced to metallic silver. Of the three Group 1 cations, only Ag^+ forms common and well-defined complex ions. For example, if $6 M$ NH_3 is added to a mixture of AgCl and Hg_2Cl_2, AgCl reacts to form a soluble complex ion; Hg_2Cl_2 does not. This difference in behavior is the basis for the separation of Ag^+ and Hg_2^{2+}. In appearing to dissolve in $6 M$ NH_3, the silver ion in AgCl actually undergoes a reaction with ammonia molecules:

$$AgCl(s) + 2NH_3 \rightarrow Ag(NH_3)_2^+ + Cl^- \qquad (4)$$

The product is the soluble, colorless complex ion diamminesilver(I).

Although silver chloride is quite insoluble in water, reaction (4) proceeds spontaneously in ammonia solution, forming the even more stable $Ag(NH_3)_2^+$ ion.

Reaction (4) is reversed in acid solution. Addition of nitric acid to a solution containing $Ag(NH_3)_2^+$ and Cl^- results in the precipitation of AgCl:

$$Ag(NH_3)_2^+ + 2H^+ + Cl^- \rightarrow AgCl(s) + 2NH_4^+ \qquad (5)$$

The reaction of $Ag(NH_3)_2^+$ with H^+ can be understood in terms of the competition between H^+ and Ag^+ for the ammonia molecules. Each of the Lewis acids, Ag^+ and H^+, tends to react with the Lewis base, NH_3. In basic solutions, the concentration of H^+ is too low to compete with Ag^+ for NH_3. However, in acid solution, protons remove NH_3 from the silver ion and Ag^+ recombines with Cl^-. The silver chloride formed is a white solid. When exposed to light for a period of time, AgCl turns gray. This results from a light-induced conversion of some of the Ag^+ ions to finely divided black particles of metallic silver.

In addition to complex-ion formation, silver ion reacts with a variety of anions to produce insoluble salts. Some important examples are AgCl (white), AgI (yellow), Ag_2S (black), Ag_2CO_3 (white), Ag_3PO_4 (yellow), Ag_2O (brown), and Ag_2CrO_4 (red). The nature of some of these silver salts is discussed in more detail in Chapter 13.

Mercury(I)

The common oxidation states of mercury are $+1$ and $+2$. The higher state, Hg^{2+}, occurs in Group 2 chemistry; the lower $+1$ state occurs in Group 1. Compounds of mercury(I) are interesting in that they involve a dimeric cation, Hg_2^{2+}. That is, mercury(I) cations are correctly represented as Hg_2^{2+} rather than Hg^+. A bond between two Hg^+ ions is not surprising. Since the outer electron configuration of atomic mercury is $5d^{10}6s^2$, that of Hg^+ must be $5d^{10}6s^1$. Sharing of $6s$ electrons could give rise to a covalent bond between the metal ions. There is good experimental evidence for the existence of Hg_2^{2+}. For example, mercury(I) compounds have no unpaired electrons (they are diamagnetic). This is consistent with Hg_2^{2+}, but not with Hg^+, which has a single, unpaired electron.

The $+1$ state lies between the 0 state of metallic mercury and the $+2$ state. Aqueous solutions of $Hg_2(NO_3)_2$ are reasonably stable. However, addition of NH_3 to Hg_2Cl_2 induces a spontaneous change from the $+1$ to the 0 and $+2$ states:

$$Hg_2Cl_2(s) \xrightarrow{NH_3} \overset{(0)}{Hg} + \overset{(+2)}{HgCl_2} \qquad (6)$$

The $HgCl_2$ produced reacts with NH_3 to give $Hg(NH_2)Cl$:

$$HgCl_2 + 2NH_3 \rightarrow Hg(NH_2)Cl(s) + NH_4^+ + Cl^- \qquad (7)$$

The amidochloride is a white solid, but it appears almost black because of the presence of finely divided mercury. The presence of Hg(I) is confirmed by the formation of a dark gray or black precipitate when 6 M NH_3 is added to mercury(I) chloride. The overall reaction for the confirmation of mercury(I) is

$$Hg_2Cl_2(s) + 2NH_3 \rightarrow Hg(NH_2)Cl(s) + Hg + NH_4^+ + Cl^- \tag{8}$$

Lead

Lead chloride is only moderately insoluble in water at room temperature. The solubility of $PbCl_2$ is greatly increased in hot water, as shown in Figure 8.1. The behavior of $PbCl_2$ can be understood in terms of a shift to the right at higher temperatures of the following equilibrium:

$$PbCl_2(s) + heat \rightleftarrows Pb^{2+} + 2Cl^- \tag{9}$$

Lead chloride may be separated from $AgCl$ and Hg_2Cl_2 by adding hot water to a mixture of the three chlorides. When the resulting solution of Pb^{2+} and Cl^- cools, solid lead chloride is formed again. Addition of K_2CrO_4 to the hot solution containing Pb^{2+} causes the formation of yellow lead chromate:

$$Pb^{2+} + CrO_4^{2-} \rightarrow PbCrO_4(s) \tag{10}$$

The formation of a yellow precipitate of $PbCrO_4$ confirms the presence of lead in the original mixture. Other aspects of the chemistry of lead are discussed in Chapter 9.

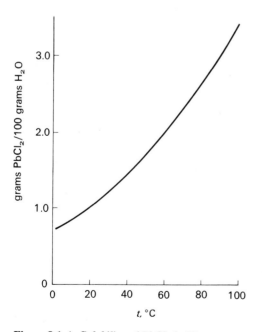

Figure 8.1 *Solubility of $PbCl_2$ in Water*

GROUP 1 DIAGRAM

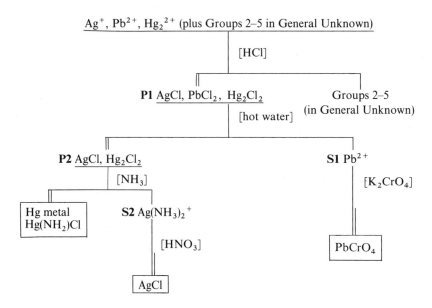

Group Diagrams

The purpose of the group diagrams is to show how the procedural steps are related to each other. Reagents to be added are indicated in brackets. In each step the resulting precipitate is shown on the left, the resulting solution on the right. The various precipitates are designated here, and in the written directions, as **P1**, **P2**, etc., the solutions as **S1**, **S2**, etc. Each precipitate or complex ion that confirms the presence of an ion is enclosed in a rectangle.

8.2 GROUP 1 PROCEDURE

(NOTE: Work on the general unknown begins on page 84.)

Group 1 Known

The Group 1 known solution consists of a mixture of $AgNO_3$, $Hg_2(NO_3)_2$, and $Pb(NO_3)_2$ in dilute HNO_3. Place 5 drops of this solution and 10 drops of distilled water into a 3-inch test tube. [If this known solution is not available in the reagent area, prepare a known mixture by placing 5 drops each of $AgNO_3$, $Hg_2(NO_3)_2$, and $Pb(NO_3)_2$ solutions in a 3-inch test tube.] Before beginning the procedure in the following paragraph, carefully review the section on techniques (page 74). Meaningful results can be obtained only by use of the proper techniques. Follow the Group 1 procedures beginning with the second paragraph below. After completing work on the known, begin work on the Group 1 unknown.

Group 1 Unknown

The instructor will provide you with an unknown that may contain any combination of $AgNO_3$, $Hg_2(NO_3)_2$, and $Pb(NO_3)_2$. Place 5 drops of this solution and 10 drops of distilled water into a 3-inch test tube. Follow the procedure for the group known below.

Add 1 drop of 6 M HCl to the solution to be tested for Group 1 cations. Stir the mixture thoroughly with a thin glass stirring rod. Test the solution with litmus or pH test paper to make sure that it is strongly acid. If it is not, add more 6 M HCl, but avoid an excess. At high chloride concentrations, $PbCl_2$ and AgCl may redissolve by forming chlorocomplexes. Heat the test tube in a hot-water bath to coagulate the precipitate. Hold the test tube under tap water until the solution is cool. Allow the mixture to stand for 2 minutes. Rub the walls of the test tube several times with a stirring rod to facilitate the precipitation of $PbCl_2$. Centrifuge. Add 1 drop of 6 M HCl to

Experimental Results

test for complete precipitation. If additional solid forms, repeat the steps above until precipitation is complete. Pour off and discard the solution.

➤ Add 8 drops of distilled water to the Group 1 precipitate, **P1**. (The symbols **P1**, **P2** and **S1**, **S2** refer to the corresponding precipitates and solutions in the Group 1 diagram.) Stir the mixture and then place the test tube in a boiling-water bath for 2 minutes. After removing the test tube from the bath, centrifuge as quickly as possible. While the mixture is still hot, pour the solution **S1** into another 3-inch test tube. Add 5 drops of distilled water to the remaining precipitate **P2**, stir, and heat in the hot-water bath. Centrifuge. Pour off and discard the wash solution. Repeat this wash procedure once more. Add 2 drops of 0.1 M K_2CrO_4 to **S1**. A yellow precipitate of $PbCrO_4$ confirms the presence of lead.

To the precipitate **P2**, add 6 drops of 6 M NH_3 and stir thoroughly. Centrifuge. The formation of a dark gray or black solid confirms the presence of Hg_2^{2+} in the sample. A small amount of a light gray solid is not a positive test. Pour the solution **S2** into another test tube. Add 6 M HNO_3 dropwise until the solution is strongly acid. Formation of a white precipitate confirms the presence of Ag^+ in the sample.

General Unknown— Group 1 Procedure

(Begin work on the general unknown here.) After you have completed the experiments on all five of the individual groups, the instructor will provide you with a general unknown. This unknown may contain any combination of the cations studied in Groups 1 through 5. Place 20 drops of the general unknown solution into a 3-inch test tube. (If the unknown is in solid form, refer to Chapter 14 for the necessary preliminary steps.) Save the remainder of the unknown; it will be used in the Groups 4 and 5 procedure.

The first step involves separating Ag^+, Hg_2^{2+}, and Pb^{2+} from the other cations in Groups 2 through 5. The general unknown is treated by the same procedure used for the Group 1 known. As you work through the steps below, refer back to your recorded observations of the Group 1 results. Experimental results for each test made on the general unknown should be recorded in the space adjacent to the procedures. Devise some method (for example, underscoring) to distinguish these results from those obtained on the Group samples. Extra space is also provided at the end of this chapter and on pages 139–142.

Add 1 drop of 6 M HCl to the 20 drops of general unknown solution. Stir the mixture and heat the test tube in a hot water bath. Next, thoroughly cool the test tube under tap water. Stir the mixture for 2 minutes, occasionally rubbing the walls of the tube with the stirring rod. (Remember that $PbCl_2$ is slow to precipitate.) If no precipitate appears, Group 1 cations are absent. In this case proceed directly to the Group 2—General Unknown Procedure on page 101. If a white solid has formed, centrifuge the mixture. Add 1 drop of 6 M HCl to test for complete precipitation. After precipitation is complete, pour the clear solution into a 3-inch test tube labeled Groups 2–5. Stopper this solution and save it for further work. Wash the Group 1 precipitate with a mixture of 1 drop of 6 M HCl and 5 drops of water. Centrifuge and discard the wash solution. Continue with the Group 1 procedure starting at the arrow on page 84.

GROUP 1—UNKNOWN REPORT SHEET

NAME_____ DATE_____

Circle the ions that are definitely present.

$$Ag^+ \quad Hg_2^{2+} \quad Pb^{2+}$$

In the space below, set up a Group 1 diagram that corresponds to the results obtained on your Group 1 unknown. Write the complete chemical formula of each precipitate and indicate its color. Write the correct formula and charge of each complex ion and indicate its color, if any.

EQUATIONS FOR REACTIONS IN THE GROUP 1 PROCEDURE

Use this space to write balanced, net-ionic equations for the reactions observed in Group 1.

Group 1: Ag^+, Pb^{2+}, and Hg_2^{2+}

Problems

1. Write a chemical equation for each reaction that silver ion undergoes in Group 1.

2. Write the chemical equation and the corresponding equilibrium-constant expression for the solubility equilibrium of Hg_2Cl_2.

3. Write the formula of a reagent that will
   a. form a precipitate with Ag^+, but not with Cu^{2+}.

   b. dissolve AgCl, but not Hg_2Cl_2.

   c. dissolve $PbCl_2$, but not Hg_2Cl_2.

4. Each solution, tested as described below, contains only one of the Group 1 cations. Indicate which ion is present in the solution. If the test results do not allow a clearcut choice to be made, indicate "insufficient information."

   a. On addition of HCl, a precipitate is formed that is insoluble in hot water but soluble in 6 M NH_3.

   b. Addition of HCl forms a precipitate that does not dissolve in 6 M NH_3.

Group 2: Hg^{2+}, As(III), Pb^{2+}, Cu^{2+}, and Cd^{2+}

CHAPTER 9

9.1 CHEMISTRY OF GROUP 2

Group Separation

The cations Hg^{2+}, As(III), Pb^{2+}, Cu^{2+}, and Cd^{2+} form highly insoluble sulfides. Addition of H_2S to a mixture of these cations in acid solution results in complete precipitation of the sulfides. Representative reactions for Hg^{2+} and Cu^{2+} are

$$Hg^{2+} + H_2S \rightarrow HgS(s) + 2H^+ \qquad (1)$$

$$Cu^{2+} + H_2S \rightarrow CuS(s) + 2H^+ \qquad (2)$$

Some of the Group 3 cations also form insoluble sulfides. However, Group 2 can be separated from Group 3 by adjusting the hydrogen-ion concentration to 0.3 M. Sulfide precipitation reactions involve a competition for the S^{2-} ion by H^+ and the cation [for example, Cu^{2+} in equation (2)]. At hydrogen-ion concentration near 0.3 M, H^+ is unable to prevent the precipitation of CuS and the other Group 2 sulfides. However, the highly hydrated Group 3 cations have a lower affinity for sulfide ion. At the 0.3 M level, hydrogen ions compete effectively for S^{2-} and prevent the formation of Group 3 sulfides. Refer to Chapter 4 for a quantitative discussion of this important separation step.

The Group 2 sulfides are all highly colored.

HgS	As_2S_3	PbS	CuS	CdS
black, brown, or red	*yellow*	*black*	*black*	*yellow*

The colors may be useful in detecting the presence of one or more sulfides. However, a given color may be misleading, especially in a mixture of sulfides. The darker sulfides completely dominate the colors of the lighter sulfides. Elemental sulfur, produced by oxidation of H_2S, should not be confused with As_2S_3 or CdS. Elemental sulfur is a light yellow solid, insoluble in water. Finely divided sulfur gives the appearance of a milky dispersion.

Because H_2S is a disagreeable and toxic substance, it is not used in these experiments in the gaseous state. Rather, it will be generated as needed by the hydrolysis of thioacetamide, CH_3CSNH_2. When heated to

80°C or more in acidic or basic solution, thioacetamide decomposes to yield H_2S.

In acidic solution:

$$\underset{\|}{CH_3}\overset{S}{C}-NH_2 + 2H_2O + H^+ \rightarrow \underset{\|}{CH_3}\overset{O}{C}-OH + NH_4^+ + H_2S \quad (3)$$

In basic solution:

$$\underset{\|}{CH_3}\overset{S}{C}-NH_2 + 2OH^- \rightarrow \underset{\|}{CH_3}\overset{O}{C}-O^- + NH_3 + HS^- \quad (4)$$

Mercury(II)

Most mercury(II) salts are insoluble in water. The common soluble salts are $Hg(NO_3)_2$ and $HgCl_2$. Mercury(II) chloride, although quite soluble, has the interesting characteristic of being only slightly dissociated into ions. Thus, $HgCl_2$ molecules, rather than Hg^{2+} and Cl^- ions, are the important species present in $HgCl_2$ solution. Both $Hg(OH)_2$ and HgS are insoluble in water. The color of HgS is either black or red, depending on the conditions. Freshly precipitated from acidic solution, HgS is black. However, in basic solution HgS is converted to a red form. Mixtures of the two appear brown. In nature, mercury is often found as the red mineral, cinnabar, having the formula HgS.

HgS is extremely insoluble in water, as judged by its $K_{sp} = 1 \times 10^{-52}$. Hydrogen ion alone, no matter how concentrated, will not dissolve HgS:

$$HgS(s) + 2H^+ \rightarrow \text{(no reaction)} \quad (5)$$

However, HgS can be dissolved by chemical reaction in two different ways. The first of these involves complex ion formation in highly basic solutions of sulfide ion:

$$HgS(s) + S^{2-} \rightarrow HgS_2^{2-} \quad (6)$$

Reaction (6) is reversed in acid solution. The solubility of HgS in basic sulfide solutions serves as a means of separating HgS (and also As_2S_3) from the other Group 2 sulfides. The second method involves both complex ion formation and oxidation of S^{2-} to free sulfur. Reaction (7) occurs when HgS is treated with a mixture of HCl and HNO_3:

$$3HgS(s) + 2NO_3^- + 12Cl^- + 8H^+ \rightarrow$$
$$3HgCl_4^{2-} + 3S + 2NO + 4H_2O \quad (7)$$

The extremely stable HgS "dissolves" only by the combined effect of the formation of $HgCl_4^{2-}$ together with oxidation of S^{2-} by NO_3^-.

Upon dilution with water, a solution of $HgCl_4^{2-}$ is converted to $HgCl_2$. The presence of mercury(II) can be confirmed by reduction of $HgCl_2$ to Hg_2Cl_2 and mercury metal. Tin(II) chloride, $SnCl_2$, is a convenient reagent for this purpose:

$$2HgCl_2 + Sn^{2+} \rightarrow Hg_2Cl_2(s) + Sn^{4+} + 2Cl^- \quad (8)$$

$$Hg_2Cl_2(s) + Sn^{2+} \rightarrow 2Hg + Sn^{4+} + 2Cl^- \qquad (9)$$

The resulting mixture will vary in color from white, if only Hg_2Cl_2 is formed, to dark gray, if metallic mercury is also formed by the action of excess $SnCl_2$.

Arsenic(III)

The important oxidation states of arsenic are $+3$ and $+5$. Only the $+3$ state occurs in the Group 2 procedure. Although As(III) is treated as a Group 2 metal ion, arsenic and its compounds possess nonmetal characteristics, also. For example, arsenic trichloride is hydrolyzed in water,

$$AsCl_3 + 3H_2O \rightarrow H_3AsO_3 + 3H^+ + 3Cl^- \qquad (10)$$

to give arsenious acid, H_3AsO_3, rather than As^{3+} ions. Like other highly charged cations, As^{3+} interacts very strongly with oxygen atoms in H_2O to form species such as H_3AsO_3. The exact nature and composition of the arsenic(III) species depends on the conditions in solution. For convenience, the more general formula, As(III), will be used.

Addition of H_2S to an As(III) solution results in the precipitation of yellow As_2S_3:

$$2As(III) + 3H_2S \rightarrow As_2S_3(s) + 6H^+ \qquad (11)$$

Since compounds of arsenic exhibit nonmetal behavior, the oxides and sulfides should be acidic in nature. Thus, As_2S_3 dissolves readily in basic solutions. If a high concentration of sulfide ions is present in a highly basic solution, reaction (12) occurs:

$$As_2S_3(s) + 3S^{2-} \rightarrow 2AsS_3^{3-} \qquad (12)$$

When As_2S_3 reacts with a mildly basic ammonia solution, in which the concentration of S^{2-} ions is low, reaction (13) probably occurs:

$$As_2S_3(s) + 6OH^- \rightarrow 2AsO_3^{3-} + 3H_2S \qquad (13)$$

These reactions provide a convenient means of separating arsenic sulfide from the other metal sulfides in Group 2.

Lead

Most lead compounds are insoluble in water. The only soluble salts encountered in Group 2 are $Pb(NO_3)_2$ and $Pb(C_2H_3O_2)_2$. Lead is included in Group 2, as well as in Group 1, because of the borderline solubility of $PbCl_2$. In a separation of the groups, lead may be found in both Group 1 and Group 2 because of incomplete precipitation of $PbCl_2$ in Group 1.

Black PbS is a basic sulfide with little tendency to dissolve in basic solutions. Thus, PbS, along the CuS and CdS, can readily be separated from mercury and arsenic. Although PbS is too insoluble to dissolve in acid solution, it will react with hot 3 M HNO_3:

$$3PbS(s) + 8H^+ + 2NO_3^- \rightarrow 3Pb^{2+} + 3S + 2NO + 4H_2O \qquad (14)$$

In equation (14) it is clear that PbS reacts with nitric acid because of the oxidation and removal of sulfide ion. Hydrogen ion alone is not effective.

Lead(II) occurs together with Cu^{2+} and Cd^{2+} in the copper subgroup of Group 2. Of the three ions, only Pb^{2+} forms an insoluble sulfate. Treatment of a mixture of these ions with sulfuric acid leads to the precipitation of $PbSO_4$. In dilute H_2SO_4 solution the reaction is

$$Pb^{2+} + SO_4^{2-} \rightarrow PbSO_4(s) \tag{15}$$

However, the solubility of $PbSO_4$ is greatly increased at high H^+ concentrations and also by the presence of NO_3^-. The method used in Group 2 to dissolve $PbSO_4$ involves treatment with ammonium acetate solution. The $C_2H_3O_2^-$ ions react to form soluble, but *nondissociated*, lead acetate:

$$PbSO_4(s) + 2C_2H_3O_2^- \rightarrow Pb(C_2H_3O_2)_2 + SO_4^{2-} \tag{16}$$

In reaction (16), $PbSO_4$ dissolves because of the formation of the non-ionized solute species, $Pb(C_2H_3O_2)_2$.

Copper(II)

Although both the +1 and +2 oxidation states of copper are important, only the +2 state is involved in Group 2 chemistry. The Cu^{2+} ion forms a number of complexes, the colors of which depend on the ligands present. Chlorocomplexes are green, the $Cu(H_2O)_6^{2+}$ hexaaquocomplex is light blue, and the familiar $Cu(NH_3)_4(H_2O)_2^{2+}$ diaquotetraammine complex ion is bright blue. The ammine complex may also be written as $Cu(NH_3)_4^{2+}$, omitting the two H_2O ligands. As discussed in Section 5.3, the bonds to the two H_2O ligands are somewhat weaker than those to NH_3.

Copper(II) hydroxide is a light blue solid, insoluble in water. Addition of dilute base to a solution of Cu^{2+} results in the precipitation of $Cu(OH)_2$:

$$Cu^{2+} + 2OH^- \rightarrow Cu(OH)_2(s) \tag{17}$$

[To simplify equation (17), the aquocomplex of copper(II) is written as Cu^{2+}.] When a fairly concentrated solution of ammonia is added to a solution of $Cu(H_2O)_6^{2+}$, the ammine complex is formed:

$$Cu(H_2O)_6^{2+} + 4NH_3 \rightarrow Cu(NH_3)_4(H_2O)_2^{2+} + 4H_2O \tag{18}$$

The important feature of reaction (18) is the replacement of four H_2O ligands by four NH_3 molecules. Thus, all ligands surrounding the Cu^{2+} ion are shown for completeness. Reaction (18) could also be written in the simplified form

$$Cu^{2+} + 4NH_3 \rightarrow Cu(NH_3)_4^{2+} \tag{19}$$

The $Cu(NH_3)_4(H_2O)_2^{2+}$ ion is stable in basic solution, but addition of acid removes the NH_3 ligands, replacing them with water molecules again:

$$Cu(NH_3)_4(H_2O)_2^{2+} + 4H_3O^+ \rightarrow Cu(H_2O)_6^{2+} + 4NH_4^+ \tag{20}$$

Copper(II) can be reduced to the +1 state with a mild reducing agent such as I^-,

$$2Cu^{2+} + 4I^- \rightarrow 2CuI(s) + I_2 \qquad (21)$$

or to the 0 state with a strong reducing agent such as sodium dithionite, $Na_2S_2O_4$:

$$Cu(NH_3)_4(H_2O)_2^{2+} + S_2O_4^{2-} \rightarrow Cu(s) + 2SO_3^{2-} + 4NH_4^+ \qquad (22)$$

Reaction (22) provides a convenient means of removing copper(II) from solution by reducing it to solid copper metal.

Cadmium

Metallic cadmium resembles zinc in appearance and reactivity. Although seldom used in the pure state, cadmium metal is a component of a variety of alloys. In solution, only the Cd^{2+} ion is of importance. Like copper, $CdCl_2$, $Cd(NO_3)_2$, and $CdSO_4$ are soluble in water, whereas $Cd(OH)_2$ and CdS are insoluble. Addition of NH_3 to a solution of Cd^{2+} produces the tetraammine complex ion:

$$Cd^{2+} + 4NH_3 \rightarrow Cd(NH_3)_4^{2+} \qquad (23)$$

Since it is more difficult to reduce Cd(II) that Cu(II), sodium dithionite does not react with $Cd(NH_3)_4^{2+}$. This difference in reactivity between cadmium and copper can be exploited to separate the two ions in basic solution. Cadmium sulfide is more stable than $Cd(NH_3)_4^{2+}$, so addition of H_2S to $Cd(NH_3)_4^{2+}$ results in the precipitation of yellow cadmium sulfide:

$$Cd(NH_3)_4^{2+} + H_2S \rightarrow CdS(s) + 2NH_4^+ + 2NH_3 \qquad (24)$$

GROUP 2 DIAGRAM

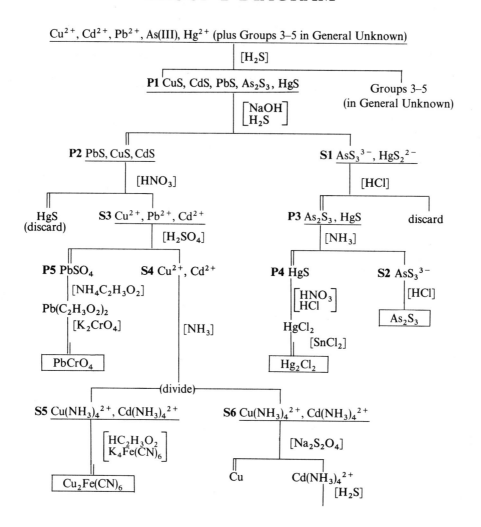

9.2 GROUP 2 PROCEDURE

Group 2 Known

The Group 2 known solution is prepared from two separate solutions. The first of these contains $Cd(NO_3)_2$, $Cu(NO_3)_2$, $Pb(NO_3)_2$, and $Hg(NO_3)_2$ in dilute HNO_3. Place 5 drops of this solution and 5 drops of distilled water into a 3-inch test tube. Add 5 drops of an As(III) solution which contains As_2O_3 dissolved in dilute HCl. Follow the Group 2 procedures beginning with the second paragraph below. (Do not begin work on Group 2 unless you have at least $\frac{1}{2}$ hour of working time available.) After completing work on the known, begin work on the Group 2 unknown.

Group 2 Unknown

The unknown may contain any combination of Group 2 cations. Place 5 drops of the unknown and 10 drops of water into a 3-inch test tube. Follow the procedure for the group known below.

Prepare a reference solution of 0.3 M HCl by *mixing* 2 drops of 2 M HCl and 11 drops of water in a 3-inch test tube. Using a stirring rod, place 1 drop of the solution on a small piece of methyl violet paper. The resulting blue-green color indicates that $[H^+]$ is approximately 0.3 M HCl. With more acidic solutions the paper turns yellow. With more basic solutions the paper turns purple. The solution to be analyzed for Group 2 ions must be adjusted to approximately 0.3 M H^+. Add 1 drop of 6 M HCl to the Group 2 solution. Stir thoroughly and test the acidity with methyl violet paper. If necessary, add either 2 M NH_3 or 2 M HCl to further adjust the acidity to the desired level. Add 7 drops of thioacetamide solution and place the test tube in a boiling-water bath. Heat for 6 minutes with occasional stirring. Cool the test tube and then centrifuge. Pour off and discard the liquid layer. Wash the precipitate **P1** with a mixture of 10 drops of distilled water and 2 drops of 3 M NH_4Cl. Centrifuge and discard the

Experimental Results

wash solution. (If you stop work at this point, cover the precipitate with 10 drops of water and 1 drop of thioacetamide. At the beginning of the next lab period, centrifuge and discard this protective solution.)

To the precipitate **P1** of Group 2 sulfides, add 10 drops of water, 6 drops of 6 M NaOH, and 6 drops of thioacetamide. (CAUTION: *Wear safety glasses. Keep solutions containing NaOH away from your face.*) Heat the mixture in a boiling-water bath for a full 3 minutes. *The open end of the tube should be pointed away from you.* Stir frequently. Cool the test tube and centrifuge. Pour off the solution **S1** into a 3-inch test tube. Wash the remaining precipitate **P2** with 10 drops of distilled water. Further procedural steps for **P2** are described on page 99. (If you do not plan to work on **P2** until the next lab period, temporarily cover **P2** with 5 drops of water and 1 drop of thioacetamide.)

Mercury–arsenic subgroup. Solution **S1** may contain arsenic and mercury in the form of the thiocomplexes, AsS_3^{3-} and HgS_2^{2-}. Add 6 M HCl to **S1**, 1 drop at a time, until the mixture is strongly acid. Stir after each drop is added and check the pH. The formation of a yellow precipitate indicates that arsenic is present. A dark precipitate suggests that mercury (and perhaps arsenic) is present. If no solid forms (except for a possible milky suspension of sulfur), arsenic and mercury are absent. Centrifuge the mixture and discard the solution. Wash **P3** with 5 drops of water. Add 6 drops of 6 M NH_3 to **P3** and stir thoroughly to dissolve As_2S_3, if present. Centrifuge. Pour off and save solution **S2**. Treat the remaining solid with another 6 drops of 6 M NH_3. Centrifuge and add the liquid layer to **S2**. If, after centrifuging, the solution is dark and cloudy, add 2 drops of 3 M NH_4Cl and centrifuge again. Add 6 M HCl to solution **S2** until it is strongly acid. The formation of a yellow precipitate of As_2S_3 confirms the presence of arsenic.

A dark precipitate **P4** which remains after the 6 M NH_3 treatment is probably

HgS. (Wear safety glasses.) Add 2 drops of 6 M HNO_3 and 5 drops of 6 M HCl to **P4**. Heat the test tube in a boiling-water bath for 3 minutes. Stir the mixture to keep the particles in contact with the solution. Continue to heat and stir after reaction occurs. Cool the test tube and then add 10 drops of water and stir. Centrifuge and discard any solid present. Add 3 drops of 0.2 M $SnCl_2$ solution. A white precipitate of Hg_2Cl_2, which may turn gray due to formation of Hg, confirms the presence of mercury. [$SnCl_2$ is susceptible to air oxidation. Before using the $SnCl_2$ reagent, check it by adding 3 drops to a solution containing 1 drop of $Hg(NO_3)_2$, 1 drop of 6 M HCl, and 5 drops of water.]

Copper subgroup. Precipitate **P2** may contain PbS, CuS, and CdS. Centrifuge and discard any liquid layer that may be mixed with **P2**. Add 6 drops of water and 6 drops of 6 M HNO_3 to **P2**. Heat the test tube in a boiling-water bath for 3 minutes. Stir the mixture to keep the particles in contact with the solution. Cool the test tube and centrifuge. Transfer the liquid layer **S3** to an evaporating dish. Any precipitate remaining should be discarded. It consists of sulfur (perhaps, dark in color) and any undissolved mercuric sulfide. However, if a large amount of **P2** still remains, repeat the treatment of **P2** with 3 drops of water and 3 drops of 6 M HNO_3. Combine the liquid layer with **S3**.

Wear safety glasses. Keep the evaporating dish away from your face during the following steps. Slowly add 3 drops of 18 M H_2SO_4 to **S3**. Set up a burner *in the hood*. Cautiously heat the evaporating dish to volatilize the HNO_3 present. HNO_3 interferes with the precipitation of $PbSO_4$ in the next step. At first, a white vapor is released. When the volume of the solution decreases to a few drops, SO_3 will be emitted as a dense, white smoke. At this point stop heating and allow the evaporating dish to cool to room temperature. When the dish is cool, *cautiously* add 8 drops of water. Allow several minutes for the lead sulfate to form. Stir the bottom and sides of the dish to contact all the material. Immediately after stirring transfer the mixture to a 3-inch test tube. Complete the transfer with 8 more drops of water. The $PbSO_4$ may be difficult to see. Proceed with the next steps, even if $PbSO_4$ is present in small amount. Centrifuge. Pour off and save solution **S4**. If there is enough precipitate **P5** remaining, wash it with 5 drops of water. Discard the wash solution. Add 8 drops of 2 M $NH_4C_2H_3O_2$ to **P5** and briefly heat and stir. If any solid remains after cooling, transfer the clear solution to another test tube. Add 1 drop of 0.1 M K_2CrO_4. Mix and centrifuge. A yellow precipitate of $PbCrO_4$ confirms the presence of lead.

Add 6 M NH_3 to **S4** dropwise until the solution is basic. Stir the solution thoroughly and make sure that it is definitely basic. A blue color at this point indicates that $Cu(NH_3)_4^{2+}$ is present. Divide the solution by placing approximately one half the solution into another test tube labeled **S6**. The remaining solution should be marked **S5**. Test solution **S5** for copper, whether or not the solution is blue. Acidify **S5** by dropwise addition of 6 M $HC_2H_3O_2$. Add 2 drops of 0.05 M $K_4Fe(CN)_6$. A cloudy, red precipitate of $Cu_2Fe(CN)_6$ confirms the presence of copper. If a white, rather than a red, precipitate is formed at this point, cadmium may be present in the sample. [Solutions of $K_4Fe(CN)_6$ are subject to air oxidation. If the test for copper is inconclusive, check the $K_4Fe(CN)_6$ by adding it directly to a dilute $Cu(NO_3)_2$ solution.]

Solution **S6** may contain a mixture of $Cu(NH_3)_4^{2+}$ and $Cd(NH_3)_4^{2+}$. Before testing for cadmium, it is necessary to remove copper, and traces of other metal ions, from solution. This can be accomplished by reduction of Cu(II) to copper metal with sodium dithionite. (The dithionite reduction should be carried out whether or not copper is present.) The steps below are to be carried out on solution **S6**. However, before committing the **S6** sample, it is recommended that you practice on trial solutions A and B. <u>Solution A</u>: Place 2 drops of $Cu(NO_3)_2$, 3 drops of water, and 3 drops of 6 M NH_3 into a 3-inch test tube. Solution B: Place 2 drops of $Cu(NO_3)_2$, 2 drops of $Cd(NO_3)_2$, 3 drops of 6 M NH_3, and 3 drops of water into a test tube. To each of the trial solutions, add a small amount of the solid dithionite, $Na_2S_2O_4$, about as much as can be placed on the extreme tip of a wood splint or spatula. Stir and then heat the tubes for 1 minute in a hot-water bath. Do not overheat. If copper is present, a red precipitate of copper metal will form. Quickly cool the tubes under tap water and then centrifuge. If a red solid forms, repeat the process with a smaller amount of $Na_2S_2O_4$. After centrifuging, pour the liquid layer into a clean test tube. Add 3 drops of thioacetamide to the liquid and heat. The appearance of a heavy, yellow precipitate confirms the presence of cadmium. (Formation of a yellow color may occur before thioacetamide is added. Presumably, this results from sulfide formed by the decomposition of $Na_2S_2O_4$.) Cadmium should be detected in trial solution B but not in A. After working on the trial solutions, repeat the above procedure on solution **S6**.

General Unknown—Group 2 Procedure

The HCl solution, resulting from the precipitation of Group 1, contains cations from Groups 2 through 5. Group 2 is precipitated as a mixture of sulfides. First, it is necessary to adjust the hydrogen-ion concentration to approximately 0.3 M. Check the acidity of the Groups 2 through 5 solution with methyl violet paper. Use 2 M NH_3 or 2 M HCl to adjust the acidity of the solution to 0.3 M. (Refer to the procedure for the Group 2 known.) Add 7 drops of thioacetamide and heat the mixture for 6 minutes in a boiling-water bath with occasional stirring. Centrifuge and pour off the liquid layer into another 3-inch test tube. Add 3 drops of thioacetamide to this solution and heat for 3 minutes. If further precipitation occurs, combine the solid with the Group 2 precipitate **P1**. Repeat this procedure, if necessary, until precipitation is complete. The solution obtained from the precipitation of Group 2 contains the cations of Groups 3 through 5. Transfer this solution to the evaporating dish. Heat the solution to the point of gentle boiling for 5 minutes. Maintain the volume of the solution by frequent addition of distilled water. The final volume should be about 15 drops. This step is important because heating removes H_2S, which will interfere in Group 3. Transfer the solution to a test tube labeled Groups 3–5 and save it for further analysis. Rinse the evaporating dish with 5 drops of water and add to the Groups 3–5 solution.

The precipitate of Group 2 sulfides **P1** should be washed twice with a mixture of 10 drops of hot water and 2 drops of 3 M NH_4Cl. Stir thoroughly during each washing. Centrifuge and discard the wash solution. Continue the Group 2 procedure starting at the arrow on page 98. (If you stop work at this point, cover the precipitate with 10 drops of water and 1 drop of thioacetamide.)

GROUP 2—UNKNOWN REPORT SHEET

NAME_____   DATE_____

Circle the ions that are definitely present.

Cu^{2+} Cd^{2+} Pb^{2+} As(III) Hg^{2+}

In the space below, set up a Group 2 diagram that corresponds to the results obtained on your Group 2 unknown. Write the complete chemical formula of each precipitate and indicate its color. Write the correct formula and charge of each complex ion and indicate its color, if any.

EQUATIONS FOR REACTIONS IN THE GROUP 2 PROCEDURE

Use this space to write balanced, net-ionic equations for the reactions observed in Group 2.

Group 2: Hg^{2+}, As(III), Pb^{2+}, Cu^{2+}, and Cd^{2+}

Problems

1. Select one of the cations Cu^{2+}, Pb^{2+}, or Hg^{2+} and write a chemical equation for each reaction that it undergoes in Group 2.

2. Write a balanced equation for each reaction. If no reaction occurs under the conditions indicated, write N.R.
   a. Lead nitrate reacts with H_2S in acid solution.

   b. Lead sulfide is heated for 3 minutes with a mixture of 2 M NaOH and thioacetamide.

   c. A solution containing the AsS_3^{3-} ion is acidified with excess HCl.

   d. Arsenic(III) sulfide dissolves in 6 M NH_3.

   e. Concentrated H_2SO_4 is strongly heated in an evaporating dish, releasing SO_3.

f. Cadmium nitrate is treated with concentrated H₂SO₄.

g. A solution containing Cu(NH₃)₄²⁺ is neutralized with 6 *M* acetic acid.

h. The presence of copper(II) is confirmed using potassium hexacyano-ferrate(II) reagent.

3. For each solution, write the formula of a test reagent that could be used to identify the underlined cation in the mixture of ions. Briefly state what the results of the test would be.
   a. <u>Cu²⁺</u> and Cd²⁺.

   b. <u>Pb²⁺</u> and Cd²⁺.

   c. <u>Sn²⁺</u> and As(III).

4. Suppose that a new cation, known to form an insoluble sulfide in acid solution, were to be analyzed according to the Group 2 procedures. What tests would have to be made on this cation to determine whether it would appear in solution **S1** or in **S3**?

Group 2: Hg^{2+}, As(III), Pb^{2+}, Cu^{2+}, and Cd^{2+}

5. Write the formula of a reagent that will
   a. form a precipitate with Cu^{2+} but not with Cd^{2+}. _____
   b. dissolve solid $CuSO_4$ but not $PbSO_4$. _____
   c. dissolve HgS but not CuS. _____
   d. dissolve CuS but not HgS. _____

6. From Group 2 chemistry, give one or more examples of the following.
   a. Strong reducing agent. _____
   b. Strong oxidizing agent. _____
   c. Species that form complex ions. _____
   d. Cations that form soluble, but nondissociated, salts. _____
   e. Cations that are not in their highest oxidation state and, therefore, can be oxidized by a suitable oxidizing agent. _____

Group 3: Fe^{3+}, Ni^{2+}, Al^{3+}, and Cr^{3+}

CHAPTER 10

10.1 CHEMISTRY OF GROUP 3

Group Separation

The four cations Fe^{3+}, Ni^{2+}, Al^{3+}, and Cr^{3+} are grouped together because they are insoluble in basic sulfide solution. In the separation of Group 3, Fe^{3+}, Al^{3+}, and Cr^{3+} are precipitated as hydroxides in an NH_3–NH_4Cl buffer. (As explained in the section concerning the chemistry of iron, either Fe^{2+} or Fe^{3+} may be present in solution, depending on the conditions. Only the equations involving Fe^{3+} are written here.)

$$Fe^{3+} + 3OH^- \rightarrow Fe(OH)_3(s) \qquad (1)$$

$$Al^{3+} + 3OH^- \rightarrow Al(OH)_3(s) \qquad (2)$$

$$Cr^{3+} + 3OH^- \rightarrow Cr(OH)_3(s) \qquad (3)$$

$$Ni(H_2O)_6^{2+} + 6NH_3 \rightarrow Ni(NH_3)_6^{2+} + 6H_2O \qquad (4)$$

The pH of an ammonia buffer solution is approximately 9, corresponding to $[OH^-] \sim 10^{-5} \, M$. Even though $[OH^-]$ is not very large, the cations are completely converted to the extremely insoluble hydroxides. In the presence of ammonia, Ni^{2+} is converted to its ammine complex. The precipitation of nickel is brought about by treatment with hydrogen sulfide:

$$Ni(NH_3)_6^{2+} + HS^- \rightarrow NiS(s) + 5NH_3 + NH_4^+ \qquad (5)$$

At a pH of 9, HS^- is the principal sulfide species.

Iron(III)–Iron(II)

In aqueous solution, both the $+2$ and $+3$ oxidation states are important. Ferrous and ferric iron frequently occur together, the relative amounts depending on the existing conditions. A reducing agent, such as H_2S, readily converts iron(III) to iron(II):

$$2Fe^{3+} + H_2S \rightarrow 2Fe^{2+} + 2H^+ + S(s) \qquad (6)$$

This reaction occurs if Fe^{3+} is present in the acid–sulfide solution for the separation of Group 2. Oxidizing agents convert iron(II) to iron (III).

Air oxidation occurs readily in basic solution:

$$4Fe(OH)_2(s) + O_2 + 2H_2O \rightarrow 4Fe(OH)_3(s) \qquad (7)$$

A light green precipitate of $Fe(OH)_2$ slowly turns red-brown as $Fe(OH)_3$ is produced.

Iron(III) exists as a six-coordinate complex ion in aqueous solution. In very acidic solution, the nearly colorless $Fe(H_2O)_6^{3+}$ ion is present. However, most solutions containing iron(III) are yellow to orange-brown in color. This is a result of the hydrolysis of hexaaquoiron(III):

$$Fe(H_2O)_6^{3+} + H_2O \rightleftarrows Fe(H_2O)_5OH^{2+} + H_3O^+ \qquad (8)$$

By gradually making a solution of iron(III) more basic, equilibrium (8) is shifted to the right. A complex series of reactions can then occur, eventually leading to a hydrated precipitate of Fe_2O_3. It will be sufficient for our purposes to summarize this change in the following way:

$$Fe(H_2O)_6^{3+} + 3OH^- \rightarrow Fe(H_2O)_3(OH)_3(s) + 3H_2O \qquad (9)$$

or

$$Fe^{3+} + 3OH^- \rightarrow Fe(OH)_3(s) \qquad (10)$$

When thiocyanate ion, SCN^-, is added to a solution containing Fe^{3+} a bright red color results. The color arises from the $Fe(H_2O)_5NCS^{2+}$ ion formed by replacement of a water molecule in the first coordination sphere of iron:

$$Fe(H_2O)_6^{3+} + SCN^- \rightleftarrows Fe(H_2O)_5NCS^{2+} + H_2O \qquad (11)$$

Since other Group 3 ions do not form a red thiocyanate complex, reaction (11) provides a sensitive test for Fe^{3+}.

Nickel

Only the +2 oxidation state of nickel occurs in aqueous solution. Thus, nickel(II) normally does not undergo oxidation–reduction reactions. $Ni(NO_3)_2$, $NiCl_2$, and $NiSO_4$ are common nickel salts, soluble in water. Nickel hydroxide is insoluble in water. Addition of strong base to a nickel(II) solution produces a green precipitate of $Ni(OH)_2$:

$$Ni^{2+} + 2OH^- \rightarrow Ni(OH)_2(s) \qquad (12)$$

The $Ni(H_2O)_6^{2+}$ ion is light green in color. As is the case for other transition metal complexes, the color observed depends on the ligands present. Addition of ammonia to a nickel(II) solution results in the formation of the blue-violet $Ni(NH_3)_6^{2+}$ ion:

$$Ni(H_2O)_6^{2+} + 6NH_3 \rightarrow Ni(NH_3)_6^{2+} + 6H_2O \qquad (13)$$

Reaction (13), in which ammonia replaces all six of the H_2O ligands, is complete only at high ammonia concentrations. In dilute solution, other complexes, containing a mixture of the two ligands, are present. The use of the equation showing complete replacement is sufficient for our purposes.

A strawberry red solid is formed when dimethylglyoxime (DMGH) is added to an ammonia solution of Ni(II):

$$Ni(NH_3)_6^{2+} + 2DMGH \rightarrow Ni(DMG)_2(s) + 2NH_4^+ + 4NH_3 \quad (14)$$

Ni(DMG)$_2$ is an insoluble coordination compound. (Its chemical structure is described on page 53.) Again, a change in ligand produces a change in color. The red solid provides a sensitive confirmatory test for the presence of Ni^{2+}.

Aluminum

The Al(H$_2$O)$_6^{3+}$ ion is colorless. Aluminum is not a transition metal, so light absorption by electron transitions among d energy levels does not occur. Since only the +3 state exists, aluminum does not undergo oxidation or reduction. In solutions that have a low concentration of hydroxide ion, for example aqueous ammonia, Al^{3+} forms insoluble Al(OH)$_3$ [equation(2)]. In strongly basic solution, for example aqueous NaOH, Al(OH)$_3$ reacts to form a soluble complex ion:

$$Al(OH)_3(s) + OH^- \rightarrow Al(OH)_4^- \quad (15)$$

Thus, aluminum exhibits amphoteric properties. The hydroxy compound Al(OH)$_3$ goes into solution by reacting with both H$^+$ and OH$^-$. A schematic representation of these changes is

$$Al^{3+} \underset{\text{addition of excess strong acid}}{\overset{\text{weak base [NH}_3\text{]}}{\rightleftarrows}} Al(OH)_3(s) \underset{\text{addition of a small amount of acid}}{\overset{\text{strong base [NaOH]}}{\rightleftarrows}} Al(OH)_4^-$$

The colorless tetrahydroxoaluminate(III) ion can be partially neutralized with acid to form Al(OH)$_3$. Addition of an excess of HCl or HNO$_3$ converts Al(OH)$_3$ to the soluble hexaaquo ion. Aluminum occurs in nature as Al$_2$O$_3$ in the ore bauxite. Al$_2$O$_3$ is the base anhydride of Al(OH)$_3$. Commercial recovery of Al$_2$O$_3$ from the ore involves the reversible formation of Al(OH)$_4^-$ described above.

Al(OH)$_3$, hydrated aluminum oxide, is a translucent, gelatinous solid, quite unlike other precipitates. Because of its very large surface area, Al(OH)$_3$ is a good chemical adsorbent. The adsorbing character of Al(OH)$_3$ provides a useful confirmatory test for aluminum. The organic dye, aluminon, is adsorbed onto the surface of Al(OH)$_3$, giving the appearance of a cherry red precipitate. Cr(OH)$_3$, which is otherwise chemically similar to Al(OH)$_3$, does not form a red precipitate with aluminon.

Chromium

The important oxidation states of chromium in aqueous solution are +3 and +6. The colors of the chromium(III) complex ions vary from violet to green, depending on the ligands present. Cr(H$_2$O)$_6^{3+}$ is violet. When one or more of the water molecules is replaced by some anion, the color is usually green. Chromium(III) closely resembles aluminum(III) in its acid–base behavior. Cr(OH)$_3$ and Al(OH)$_3$ are amphoteric; that is, they react either as an acid or as a base. Addition of a weak base, such as

NH_3, to Cr^{3+} produces a gray-green precipitate of $Cr(OH)_3$ [equation (3)]. An excess of a strong base converts this precipitate to the soluble complex, $Cr(OH)_4^-$, which is bright green in color:

$$Cr(OH)_3(s) + OH^- \rightarrow Cr(OH)_4^- \qquad (16)$$

The exact nature of the basic complex ion is unclear. The composition may also be represented as $CrO_2^- \cdot xH_2O$ or $Cr(H_2O)_2(OH)_4^-$. It has proved very difficult to devise experiments that can distinguish which forms are the correct ones.

The general chemical characteristics of chromium in solution are outlined in Figure 10.1. As shown in Figure 10.1, chromium(III) can be oxidized in basic solution to chromium(VI). An effective oxidizing agent for this purpose is hydrogen peroxide:

$$2Cr(OH)_4^- + 3H_2O_2 + 2OH^- \xrightarrow{base} 2CrO_4^{2-} + 8H_2O \qquad (17)$$

The $+6$ state is the highest expected for chromium, consistent with the six valence electrons in the $4s^1 3d^5$ atomic structure of chromium. The chromate ion, CrO_4^{2-}, is characterized by its bright yellow color. The presence of CrO_4^{2-} is confirmed by the formation of a yellow precipitate of lead chromate:

$$Pb^{2+} + CrO_4^{2-} \rightarrow PbCrO_4(s) \qquad (18)$$

Addition of acid to a CrO_4^{2-} solution results in the formation of $HCrO_4^-$ and $Cr_2O_7^{2-}$:

$$H^+ + CrO_4^{2-} \rightleftarrows HCrO_4^- \qquad (19)$$

$$2HCrO_4^- \rightleftarrows Cr_2O_7^{2-} + H_2O \qquad (20)$$

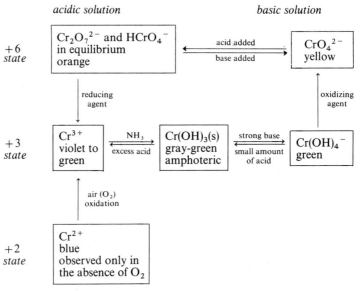

Figure 10.1 | *Chromium Species in Water Solution*

The dichromate ion, $Cr_2O_7^{2-}$, is orange. In this dimeric species, the chromium atoms are linked by a bridging oxygen atom:

$$\left[\begin{array}{c} O \\ | \\ O \end{array} \!\!\! \begin{array}{c} O \\ \diagdown \\ Cr \\ \diagup \\ O \end{array} \!\!\! O \!\!\! \begin{array}{c} \diagdown \\ Cr \\ \diagup \end{array} \!\!\! \begin{array}{c} O \\ \\ O \end{array} \right]^{2-}$$

(oxygen atoms arranged in a tetrahedron around each Cr)

In acid solution, chromium(VI) is a strong oxidizing agent. Reducing species, such as H_2S, convert chromium from the $+6$ to the $+3$ state:

$$Cr_2O_7^{2-} + 3H_2S + 8H^+ \rightarrow 2Cr^{3+} + 3S + 7H_2O \qquad (21)$$

Hydrogen peroxide also will reduce dichromate in acid solution (an undesirable reaction if it should occur in the Group 3 procedure):

$$Cr_2O_7^{2-} + 3H_2O_2 + 8H^+ \xrightarrow{\text{acid}} 2Cr^{3+} + 3O_2 + 7H_2O \qquad (22)$$

Comparing equations (17) and (22), it is clear that H_2O_2 can function both as an oxidizing agent (basic solution) and as a reducing agent (acidic solution). This dual behavior arises from the intermediate -1 oxidation state of oxygen in H_2O_2.

GROUP 3 DIAGRAM

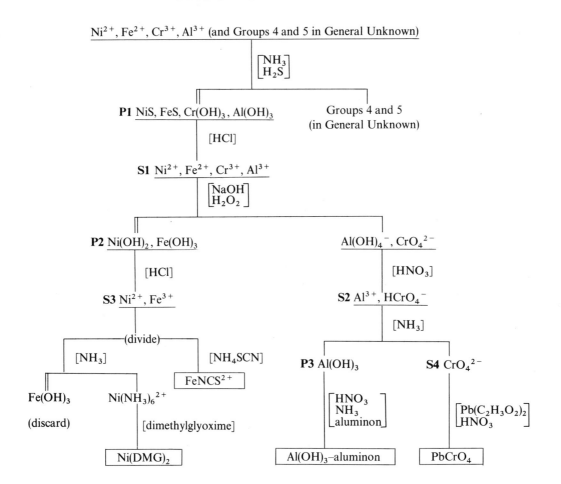

10.2 GROUP 3 PROCEDURE

Group 3 Known

The Group 3 known solution contains a mixture of $Ni(NO_3)_2$, $Fe(NO_3)_3$, $Cr(NO_3)_3$, and $Al(NO_3)_3$ in dilute nitric acid. Place 5 drops of the Group 3 known and 10 drops of distilled water into a 3-inch test tube. [If this known solution is not available in the reagent area, prepare a mixture containing 4 drops each of $Ni(NO_3)_2$, $Fe(NO_3)_3$, $Cr(NO_3)_3$, and $Al(NO_3)_3$ solutions in a 3-inch test tube.] Follow the Group 3 procedures beginning at the arrow below. (Do not begin work on Group 3 unless you have at least $\frac{1}{2}$ hour.) After completing work on the known, begin work on the Group 3 unknown.

Group 3 Unknown

The Group 3 unknown may contain any combination of the Group 3 cations. Place 5 drops of the unknown and 10 drops of distilled water into a 3-inch test tube. Follow the procedure for the group known beginning at the arrow below.

➡ To the mixture of Group 3 cations add 2 drops of 6 M NaOH. (CAUTION: *Wear safety glasses. Keep solutions containing NaOH away from your face.*) Stir the mixture and check the pH. If the mixture is still acidic, add more NaOH until it is strongly basic. Centrifuge and then add 1 drop of 6 M NaOH to test for complete precipitation. Repeat this process if necessary until precipitation is complete. (Do not stop here. Continue work until the NaOH solutions have been neutralized.)

Add 2 drops of 3 percent H_2O_2 to the mixture. (Test the H_2O_2 reagent before using it. Place a very small amount of solid MnO_2 into a test tube and add 1 drop of H_2O_2. Oxygen gas should bubble off vigorously. If it does not, ask your instructor to replace the H_2O_2.) Stir the mixture thoroughly. Heat in a hot-water bath for 2 minutes. (Keep the open end of the test tube pointed away from yourself and

Experimental Results

others.) Cool the test tube, add 2 more drops of H_2O_2, and stir thoroughly. Place the test tube in a boiling-water bath for *at least 8 minutes* to decompose all the excess H_2O_2. Stir the mixture occasionally and continue to heat until large numbers of oxygen gas bubbles are no longer released. Centrifuge. Transfer the solution, which may contain $Al(OH)_4^-$ and CrO_4^{2-}, to another test tube. Add 6 M HNO_3, stirring after each drop, until the solution is acidic. Label the test tube **S2** and save it for later steps.

The hydroxide precipitate, **P2**, should be washed twice with a mixture of 6 drops of water and 2 drops of 6 M NaOH. Discard the wash solution. Add 10 drops of water to **P2** and then add 6 M HCl dropwise until the mixture is strongly acid. The precipitate should dissolve, although several minutes may be required for the last particles to react.

The HCl solution **S3** contains Fe^{3+} and Ni^{2+}, if these ions were present in the original sample. Pour half of **S3** into another 3-inch test tube. Test one of the tubes for nickel by adding 4 drops of 6 M NH_3. The mixture should have a strong odor of ammonia and test basic. If this is not the case, add more 6 M NH_3. Centrifuge and discard any $Fe(OH)_3$ precipitate present. If nickel is present in the solution, the pale violet $Ni(NH_3)_6^{2+}$ ion may be visible. Add 2 drops of dimethylglyoxime solution and centrifuge. A strawberry red precipitate confirms the presence of nickel. Test the second test tube containing **S3** for iron by adding 2 drops of 0.1 M NH_4SCN. A bright red color, due to $Fe(H_2O)_5NCS^{2+}$, confirms the presence of iron.

Solution **S2** contains aluminum and chromium if these ions were present in the original sample. Add 1 drop of 6 M NH_3 to the solution and stir thoroughly. Repeat this process until the mixture is definitely basic. $Al(OH)_3$ forms slowly as a translucent, jelly-like precipitate. Centrifuge. Pour the resulting solution **S4** into another test tube. A yellow color indicates that CrO_4^{2-} is present. Acidify **S4** with a minimum amount of 6 M HNO_3. Add 2 drops of 0.10 M

$Pb(C_2H_3O_2)_2$. The formation of yellow $PbCrO_4$ confirms the presence of CrO_4^{2-} in **S4** and, therefore, of Cr^{3+} in the original sample.

The precipitate **P3** should be washed with 10 drops of hot water, stirring thoroughly to remove any remaining CrO_4^{2-}. Centrifuge and discard the wash solution. (Repeat the washing procedure if a yellow color due to CrO_4^{2-} still remains.) The $Al(OH)_3$ should be very light in color. If the precipitate is gray-green, $Cr(OH)_3$ is also present and must be removed by repetition of the $NaOH-H_2O_2$ treatment.

Add 2 drops of water to the precipitate **P3**, and then add $1 M$ HNO_3 until the precipitate dissolves. Add 2 drops of aluminon dye solution and 1 drop of $6 M$ NH_3. If, after stirring, the mixture is not basic, add 1 or more drops of $6 M$ NH_3 until the mixture is basic. Centrifuge. The presence of a red solid, aluminon dye absorbed on $Al(OH)_3$, confirms the presence of aluminum. [It is recommended that the steps in this paragraph be run in parallel with a 1-drop sample of $Al(NO_3)_3$ from the reagent shelf.]

General Unknown—Group 3 Procedure

The acid–sulfide solution, resulting from precipitation of Group 2, contains cations from Groups 3, 4, and 5. Group 3 is to be precipitated as a mixture of hydroxides and sulfides. It is desirable to precipitate the hydroxides first, then the sulfides in a second step. By so doing, the nature of the hydroxides present can be examined separately. If a precipitate has formed in the Groups 3–5 solution, centrifuge and discard it. Add $6 M$ NH_3 drop by drop until the mixture is basic. Stir the mixture and check the pH after each drop is added. When the mixture is basic, add 1 more drop of $6 M$ NH_3 and 1 drop of $3 M$ NH_4Cl. Observe the nature and color of the precipitate, which may provide a clue to the cations present.

Add 7 drops of thioacetamide and place the test tube in a boiling-water bath for 3 minutes. Note any color change. Centrifuge. Test for complete precipitation of Group 3 by adding 1 drop of 6 M NH_3 and 2 drops of thioacetamide. Use a stirring rod to gently mix the solution without disturbing the solid. Heat in the water bath. When the precipitation is complete, centrifuge and pour the solution into an evaporating dish. Acidify the solution with 6 M $HC_2H_3O_2$. Heat the evaporating dish for 4 minutes to expel H_2S. Maintain an approximately constant volume of 15 drops by addition of water. It is important that H_2S be removed because the H_2S may be slowly air-oxidized to sulfate ion. Transfer the solution to a test tube labeled Groups 4 and 5. Rinse the evaporating dish with several drops of water and add to the Groups 4 and 5 solution. Save this solution for later analysis.

The mixture of Group 3 precipitates **P1** should be washed with a mixture of 15 drops of hot water and 1 drop of 3 M NH_4Cl. Centrifuge and discard the wash solution. To the Group 3 precipitate, add 2 drops of 6 M HCl. Stir the mixture and check the pH. If the mixture is not strongly acidic, add 1 or more drops of 6 M HCl as required. Heat the test tube for 2 minutes in a boiling-water bath. If no black precipitate remains, follow the procedure in the next paragraph. If a sizable quantity of a black precipitate remains, nickel sulfide probably is present. In this case, cool the test tube under tap water and add 2 drops of 6 M HNO_3. Heat the test tube again in a boiling-water bath. A vigorous reaction should occur in which most of the black solid disappears. Disregard any black particles on the surface (which probably are sulfur, colored by NiS). If the black solid has not reacted, repeat the nitric acid treatment.

Transfer the mixture to an evaporating dish, using 5 drops of water to rinse the test tube. Gently boil the mixture for 3 minutes to remove H_2S. Maintain constant volume by liberal addition of hot water.

After the dish has cooled, carefully transfer the mixture to a 3-inch test tube. Rinse the evaporating dish with several drops of water. If any solid is present, centrifuge and discard it. At this point the solution **S1** contains the Group 3 ions present in the general unknown. Continue with the Group 3 known procedure beginning at the arrow on page 115.

GROUP 3—UNKNOWN REPORT SHEET

NAME_____   DATE_____

Circle the ions that are definitely present.

Ni^{2+} Fe^{3+} Cr^{3+} Al^{3+}

In the space below, set up a Group 3 diagram that corresponds to the results obtained on your Group 3 unknown. Write the complete chemical formula for each precipitate and indicate its color. Write the correct formula and charge of each complex ion and indicate its color, if any.

EQUATIONS FOR REACTIONS IN THE GROUP 3 PROCEDURE

Use this space to write balanced, net-ionic equations for the reactions observed in Group 3.

Group 3: Fe^{3+}, Ni^{2+}, Al^{3+}, and Cr^{3+}

Problems

1. In the general unknown, nickel and iron are present as Ni^{2+} and Fe^{2+} after being separated from the precipitate of Group 2 sulfides. Write equations for the sequential steps indicated below, showing the reactions that nickel and iron undergo in Group 3.

   a. A mixture of Ni^{2+} and Fe^{2+} is made basic with 6 M NH_3 (sulfide is absent).

   b. The products of part (*a*) react with H_2S.

   c. The resulting mixture of sulfides is acidified with 6 M HCl.

   d. NaOH is added to Ni^{2+} and Fe^{2+}.

   e. The mixture of hydroxides from part (*d*) is heated with hydrogen peroxide.

   f. The products resulting from part (*e*) are acidified with 6 M HCl.

2. What differences in chemical behavior make it possible to
   a. separate Al^{3+} and Fe^{3+} in basic solution?

b. separate Fe^{3+} and Ni^{2+} in ammonia solution?

c. separate aluminum(III) and chromium(III) in basic solution?

3. Explain why chromium, but not aluminum, can exist in an oxidation state higher than $+3$.

4. What coordination complexes are important in the separation and identification of nickel(II)?

5. Write equations for the following reactions, illustrating the behavior of chromium in aqueous solution. Indicate the color of each chromium species.
   a. The weak base NH_3 is added to a solution containing Cr^{3+}.

   b. The strong base NaOH is added to a solution containing Cr^{3+}.

   c. Chromium(III) hydroxide reacts with acid.

   d. Chromium metal reacts with HCl, in the absence of oxygen, to give the hexaaquochromium(II) ion.

Group 3: Fe^{3+}, Ni^{2+}, Al^{3+}, and Cr^{3+}

    e. Hexaaquochromium(II) is oxidized by O_2 in acid solution.

    f. A strong base is added to a solution containing $Cr_2O_7^{2-}$.

6. Make sketches of the structures of the complex ions.

$Ni(NH_3)_6^{2+}$ $Cr(H_2O)_2(OH)_4^-$ (trans form)

Groups 4 and 5: Ba^{2+}, Ca^{2+} and Na^+, NH_4^+

CHAPTER 11

11.1 CHEMISTRY OF GROUPS 4 AND 5

Separation of the Groups

Because the chlorides, hydroxides, and sulfides of Ba^{2+}, Ca^{2+}, Na^+, and NH_4^+ are water soluble, these cations remain after separation of Groups 1, 2, and 3. Barium ion and calcium ion make up Group 4. Sodium ion and ammonium ion are unique in that they do not form precipitates with any common anion. Thus, Na^+ and NH_4^+ are placed into a separate group, Group 5. Barium and calcium react with carbonate ion in basic solution to form insoluble carbonates:

$$Ba^{2+} + CO_3^{2-} \rightarrow BaCO_3(s) \tag{1}$$

$$Ca^{2+} + CO_3^{2-} \rightarrow CaCO_3(s) \tag{2}$$

These reactions provide the basis for the separation of Group 4 from Group 5. The solid carbonates dissolve readily in HCl:

$$BaCO_3(s) + 2H^+ \rightarrow Ba^{2+} + CO_2 + H_2O \tag{3}$$

$$CaCO_3(s) + 2H^+ \rightarrow Ca^{2+} + CO_2 + H_2O \tag{4}$$

Barium and Calcium

These two elements are both members of group IIA in the periodic table. Accordingly, they exist exclusively as dipositive ions in solution. Ba^{2+} and Ca^{2+} are quite similar in chemical behavior. Each forms insoluble carbonate (CO_3^{2-}), phosphate (PO_4^{3-}), chromate (CrO_4^{2-}), and sulfate (SO_4^{2-}) salts. Note that each of these anions is basic (that is, the corresponding acids, HCO_3^-, $H_2PO_4^-$, etc., are incompletely dissociated). Sulfate, although not a very strong base, does combine with protons, since $K_{dissoc} = 1.2 \times 10^{-2}$ for HSO_4^-. The basicity of these anions is significant because it means that the solubility of Ba^{2+} and Ca^{2+} salts will depend on the pH of the solution. Barium and calcium are separated in the Groups 4 and 5 procedure based on the difference in solubilities of the chromate salts:

$$BaCrO_4(s) \rightleftharpoons Ba^{2+} + CrO_4^{2-} \quad (K_{sp} = 1 \times 10^{-10}) \tag{5}$$

$$CaCrO_4(s) \rightleftharpoons Ca^{2+} + CrO_4^{2-} \quad (K_{sp} = 7 \times 10^{-4}) \tag{6}$$

$$HCrO_4^- \rightleftharpoons H^+ + CrO_4^{2-} \qquad (K = 3 \times 10^{-7}) \qquad (7)$$

As shown in equilibrium (7), H^+ competes rather effectively for CrO_4^{2-} ions. For example, at $[H^+] \sim 3 \times 10^{-5}\ M$ (used in the Groups 4 and 5 procedure), only about 1 percent of the total chromate is present as CrO_4^{2-}. The remaining chromate is in the $HCrO_4^-$ form:

$$K = 3 \times 10^{-7} = \frac{[H^+][CrO_4^{2-}]}{[HCrO_4^-]} = \frac{(3 \times 10^{-5})[CrO_4^{2-}]}{[HCrO_4^-]} \qquad (8)$$

$$\frac{[CrO_4^{2-}]}{[HCrO_4^-]} = 1 \times 10^{-2} \qquad (9)$$

At pH ~ 5, the concentration of CrO_4^{2-} is too low for precipitation of $CaCrO_4$ to occur. However, Ba^{2+} competes strongly for CrO_4^{2-} ions and $BaCrO_4$ precipitates:

$$Ba^{2+} + CrO_4^{2-} \xrightarrow{pH\ \sim\ 5} BaCrO_4(s) \qquad (10)$$

$$Ca^{2+} + CrO_4^{2-} \xrightarrow{pH\ \sim\ 5} \text{(no reaction)} \qquad (11)$$

The formation of light yellow $BaCrO_4$ can be used to identify barium. Another characteristic reaction of Ba^{2+} involves formation of the acid-insoluble sulfate salt:

$$Ba^{2+} + SO_4^{2-} \rightarrow BaSO_4(s) \qquad (12)$$

Calcium ion may be precipitated from solution with any one of a number of anions. However, small amounts of Ba^{2+} in a Ca^{2+} solution might also precipitate and thereby interfere with the identification of Ca^{2+}. To minimize this possibility, oxalate ion, $C_2O_4^{2-}$, is used to precipitate Ca^{2+}:

$$Ca^{2+} + C_2O_4^{2-} \rightarrow CaC_2O_4(s) \qquad (13)$$

Calcium oxalate is more insoluble than is barium oxalate. Since CaC_2O_4 is the salt of a weak acid, CaC_2O_4 dissolves in HCl:

$$CaC_2O_4(s) + 2H^+ \rightarrow Ca^{2+} + H_2C_2O_4 \qquad (14)$$

Sodium

Na^+ is one of the least reactive cations in aqueous solution. It does not form a precipitate with any ordinary anion. It has a negligible tendency to form complex ions and does not undergo oxidation or reduction in water. Because of its presence in living cells, sodium ion is biologically important. The chemical inertness of Na^+ makes it important to the chemist as well. Addition of an anion as a sodium salt ensures that the chemical characteristics of the solution arise primarily from the anion and not from the sodium cation. Direct use of this idea is made in Part III.

The presence of sodium ion in solution can be detected by a flame ionization test. When sodium salts are energized in a high-temperature flame, a bright yellow color is imparted to the flame. Electrons, present in an excited state of the sodium atoms, emit energy in the yellow region of the visible spectrum. The yellow color of sodium is unique among the common cations. Calcium ion gives an orange flame, potassium ion a pale violet flame, and barium and copper green flames. However, none of these are as strong or as persistent as the yellow sodium flame. In a mixture, the sodium yellow usually obscures any other colors present. The measurement of light emission in a flame can be put on a quantitative basis by the technique of flame photometry. With this technique, the wavelength and intensity of visible light can be measured precisely.

Ammonium

Ammonium ion is an extremely important cation in aqueous solution. Like sodium, NH_4^+ is not very reactive in water solution. All ammonium salts are soluble in water and NH_4^+ can be oxidized only with difficulty. The NH_4^+ ion possesses a tetrahedral structure:

$$\left[\begin{array}{c} H \\ | \\ H \cdots N \diagdown H \\ | \\ H \end{array} \right]^+$$

The N—H bonds are strong, so dissociation of NH_4^+ does not occur readily. However, in basic solution, OH^- removes a proton from NH_4^+, releasing molecular ammonia, NH_3:

$$NH_4^+ + OH^- \rightleftarrows NH_3 + H_2O \qquad (15)$$

If the concentration of NH_3 is sufficiently high, some of the NH_3 can escape from solution. Reaction (15) serves as the basis for the identification of NH_4^+ by the formation of NH_3. Equilibrium reaction (15) is just the reverse of the base dissociation of NH_3 in water.

GROUPS 4 AND 5 DIAGRAM

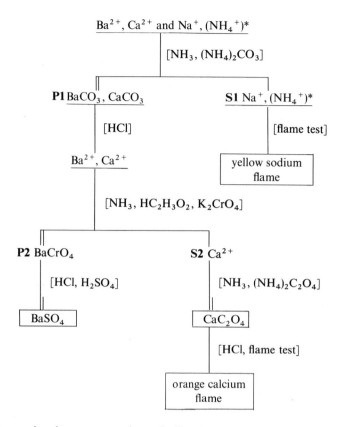

* Ammonium ion tests are to be made directly on the *original unknown mixture*.

11.2 GROUPS 4 AND 5 PROCEDURE

Ammonium-Ion Test

The first step in the Groups 4 and 5 procedure is to test for NH_4^+. This test must be made on the original sample before any reagents are added. This is necessary because reagents containing ammonium ion are added in the procedures that follow. (On the general unknown, the NH_4^+ test must be made directly on the original mixture obtained from the instructor.) Place 5 drops of the solution to be tested (see below) into the bottom of a clean, dry evaporating dish. Moisten a 1-inch strip of neutral litmus or pH test paper. Place the paper on the bottom surface of a clean watch glass. Add 5 drops of 6 M NaOH directly onto the sample in the dish and quickly place the watch glass over the evaporating dish. If NH_4^+ is present, NH_3 will be generated and partially absorbed by the test paper. A characteristic basic color, developed within 15 to 30 seconds, in the test paper confirms the presence of NH_4^+ in the sample. It may also be possible to detect the odor of ammonia. (Keep your face away from the NaOH solution. Wear safety glasses.)

Groups 4 and 5 Known

The Groups 4 and 5 known solution contains a mixture of $Ba(NO_3)_2$, $Ca(NO_3)_2$, $NaNO_3$, and NH_4NO_3. Begin by carrying out the NH_4^+ test above. Next, the Group 4 cations, Ba^{2+} and Ca^{2+}, are to be separated from the soluble Group 5 ions, Na^+ and NH_4^+, by precipitation as carbonates. Place 5 drops of the Groups 4 and 5 known, 5 drops of 3 M NH_4Cl, and 5 drops of distilled water into a 3-inch test tube. [If the known solution is not available in the reagent area, prepare a mixture containing 4 drops each of $Ba(NO_3)_2$, $Ca(NO_3)_2$, $NaNO_3$, and NH_4NO_3 solutions in a 3-inch test tube.] Follow the Groups 4 and 5 procedure beginning at the arrow below.

Experimental Results

Groups 4 and 5 Unknown (*Optional*)

After completing the Groups 4 and 5 known, consult with the instructor concerning the assignment of this unknown. If no Groups 4 and 5 unknown is assigned, proceed directly to the procedures for the general unknown, page 84. If the Groups 4 and 5 unknown is assigned, it may contain any combination of the ions Ba^{2+}, Ca^{2+}, Na^+, and NH_4^+. Place 5 drops of the unknown and 10 drops of water into a 3-inch test tube. Follow the procedure for the known beginning at the arrow below.

→ To the Groups 4 and 5 solution, add $6M$ NH_3 dropwise until the mixture is basic. Then add 3 more drops of $6\,M$ NH_3 and 6 drops of $2\,M$ $(NH_4)_2CO_3$ solution. Place the test tube in an approximately 60°C water bath for 3 minutes. Stir occasionally. Cool the test tube and centrifuge. Pour the solution **S1** into another test tube. Test **S1** for the presence of Na^+ by carrying out a flame test as described below. A genuine sodium sample will impart a solid yellow color to the flame which should persist for 3 or 4 seconds. For comparison, place 10 drops of $0.1\,M$ $NaNO_3$ into another test tube and carry out the flame test on this reference. When working on the general unknown, the sodium flame test should be made on the original mixture as well as on **S1** in Group 5. However, in so doing, care must be taken not to mistake the orange flame of calcium (if Ca^{2+} is present) as a positive test for sodium.

Flame-Test Procedure

Fill a 3-inch test tube one-half full of $12\,M$ HCl. Adjust a Bunsen burner to give a hot, all-blue flame. Clean the flame test wire and heat one end of it in the flame until the wire glows. Quickly plunge the end of the wire into the test tube containing $12\,M$ HCl. Repeat these steps until the wire no longer imparts an appreciable color to the flame. Dip the wire into the solution to be tested and then place it in the flame. Repeat this step several times to be certain of the color of the flame. For comparison, carry out the flame test on a sample known to contain the ion which is being tested for. (If difficulty in cleaning the test wire is encountered, replace the 12 M HCl solution, which may be contaminated. Also, check for and remove any particles adhering to the test wire.)

Wash the remaining precipitate **P1** twice with 8-drop portions of water. Discard the wash solutions. Add 2 drops of 2 M HCl to precipitate **P1**. If, after stirring, the solid has not completely dissolved, add one or more drops of 2 M HCl. Add 5 drops of water and then make the solution weakly basic by the addition of 6 M NH_3. Then, add 1 or more drops of 6 M $HC_2H_3O_2$ until the solution is weakly acidic. Add one extra drop of 6 M $HC_2H_3O_2$ and place the test tube in a hot-water bath. After the solution is warm, add 1 drop of 0.1 M K_2CrO_4. If a yellow precipitate of $BaCrO_4$ appears, add 5 more drops of 0.1 M K_2CrO_4 and stir thoroughly. Heat the mixture for 2 minutes, then cool and centrifuge. If the solution at this point is not distinctly yellow, add 2 more drops of the chromate solution to check for complete precipitation. Heat and centrifuge as before. Pour off the solution **S2** into another test tube. Wash the remaining precipitate **P2** with water, discarding the wash solution. Add 2 drops of 6 M HCl to **P2** and heat and stir in a hot-water bath until the solid dissolves. Cool the test tube and add 10 drops of water and 1 drop of 18 M H_2SO_4. Centrifuge. A heavy white precipitate of $BaSO_4$ confirms the presence of barium in the mixture.

Solution **S2** may contain calcium ion. Add 6 M NH_3 to **S2** until the solution is basic. Add 3 drops of 0.5 M $(NH_4)_2C_2O_4$ solution. Stir and check to be certain that the solution is basic. Allow several minutes for complete precipitation. A white precipitate of CaC_2O_4 confirms the presence of calcium. Centrifuge and discard the liquid layer. Wash the solid with a mixture of 1 drop of 0.5 M $(NH_4)_2C_2O_4$ and 10 drops of water. Discard the wash solution. Add 2 drops of 6 M HCl and heat until the solid dissolves.

Carry out a flame test on this solution as described previously for sodium. An orange flame is characteristic of calcium. For comparison, carry out a flame test on a sample known to contain Ca^{2+}. Add 2 drops of $Ca(NO_3)_2$ solution to another 3-inch test tube. Convert the Ca^{2+} to CaC_2O_4 as described earlier in this paragraph. Carry out the flame test on this known CaC_2O_4 sample.

General Unknown—Groups 4 and 5 Procedure

After separation of Group 3, the Group 4 and 5 cations remain dissolved in an acetic acid–ammonium acetate solution. The volume of this solution should be adjusted to approximately 15 drops. (Use 15 drops of water in a separate 3-inch test tube for comparison.) This can be accomplished by heating to evaporate some of the solvent in an evaporating dish, or by addition of water as required. Centrifuge and discard any precipitate present. Continue with the Groups 4 and 5 known procedure starting at the arrow on page 132. Note that the NH_4^+ test, page 131, must also be carried out, using the original unknown sample.

GROUPS 4 AND 5—UNKNOWN REPORT SHEET
(*Optional*)

NAME_____ DATE_____

Circle the ions that are definitely present.

$$Ba^{2+} \quad Ca^{2+} \quad Na^+ \quad NH_4^+$$

In the space below, set up a Groups 4 and 5 diagram that corresponds to the results obtained on your Groups 4 and 5 unknown. Write the complete chemical formula of each precipitate and indicate its color. Write the correct formula and charge of each complex ion and indicate its color, if any.

EQUATIONS FOR REACTIONS IN THE GROUPS 4–5 PROCEDURE

Use this space to write balanced, net-ionic equations for the reactions observed in Groups 4 and 5.

GENERAL UNKNOWN DATA SHEET

Use this space for additional experiment results obtained on the general unknown.

GENERAL UNKNOWN DATA SHEET

GENERAL UNKNOWN—REPORT SHEET

NAME_____ DATE_____

Circle the ions that are definitely present.

Ag^+	Hg_2^{2+}	Pb^{2+}	
Cu^{2+}	Cd^{2+}	$As(III)$	Hg^{2+}
Ni^{2+}	Fe^{3+}	Cr^{3+}	Al^{3+}
Ba^{2+}	Ca^{2+}	NH_4^+	Na^+

Problems

1. Write an equation for each reaction that barium undergoes in Group 4.

2. In the Group 4 procedure, the mixture of Ca^{2+} and Ba^{2+}, resulting from acidification of **P1**, is treated first with NH_3 and then with $HC_2H_3O_2$. What species are present in the resulting solution? What is the approximate pH of this buffer system?

3. Write the formula of a reagent that could be used to distinguish between the solids in each pair below. State what the results of the test would be.
   a. $(NH_4)_2SO_4$ and $BaSO_4$

   b. $BaCO_3$ and $BaSO_4$

   c. $CaCrO_4$ and $BaCrO_4$

PART III: Anion Chemistry and Procedures

Guide to Experimental Work

CHAPTER 12

12.1 THE FIRST LABORATORY PERIOD

Experimental techniques for anion chemistry are summarized in this chapter. Read through this material before beginning the anion experiments in Chapter 13. In particular, Sections 12.1(5), 12.3, and 12.5 contain information specific for the anion procedures and should be examined now, even if you have already completed the cation experiments in Part II.

1. *Equipment.* If this is your first period in laboratory, check the equipment assigned to you. Use the check list of equipment provided by the instructor. Report to the instructor any missing or damaged items. They will be replaced without charge during the first period. (A recommended list of equipment and supply items is shown on page 188.)
2. *Supplies.* Obtain a polyethylene wash bottle from the stockroom (optional.) Also, from the stockroom, obtain a package of supplies for the cation–anion experiments. This package contains expendable items not found in the lab desk.
3. *Stirring rods.* If thin (3 mm) glass stirring rods are not available in your desk, several of them should be made from solid rod obtained in the stockroom. The directions for making stirring rods appear on page 71.
4. *Droppers.* Several glass droppers with rubber bulbs will be needed for transferring solutions. If glass droppers are not available, they can be constructed as indicated on page 71.
5. *Desk set of reagents.* Your desk may contain a set of small bottles fitted with glass droppers. These bottles are useful for dispensing frequently used reagent solutions. If an individual desk set is not available, all necessary solutions will be found in the reagent area. (In this case, omit the steps below.)

   Clean the small reagent bottles. Discard any old solutions found in the bottles. Rinse the bottles thoroughly with distilled water. Replace any of the bottles or droppers that cannot be cleaned satisfactorily. Unless the labels are of a permanent type, remove all labels with hot water. Label the bottles with the concentration and formula of the reagent solutions listed below.

$1\,M$ HNO$_3$ $0.1\,M$ AgNO$_3$ $0.1\,M$ BaCl$_2$ KI/I$_2$

From supply bottles in the reagent area, fill each of the four bottles one-half to two-thirds full. Keep the dropper tops screwed in place except during use.

6. *Preliminary information.* Before starting experimental work, read the important preliminary sections in the remainder of this chapter.
7. *First experiments.* The procedures for the anion experiments begin in Chapter 13.

12.2 | LABORATORY SAFETY

For your own safety, please read the brief section on safety rules, pages 72–73, before undertaking any work in the laboratory.

12.3 | REAGENTS

All the reagent solutions required for the experiments are kept in the reagent supply area. These solutions should be used to fill your desk set of dropper bottles, if these bottles are available. It is important that all reagent bottles remain free of contamination. Please help maintain the integrity and availability of the reagents by observing the following rules.

1. Leave laboratory reagent bottles on the shelves and in proper order. Do not take them to your desk.
2. Put dropper tops back onto the correct bottle immediately after use. Notify the instructor if the dropper tops get mixed.
3. Do not pour unused reagent back into the supply bottle.
4. Do not let the reagents get up into the rubber bulbs in the dropper bottles. Dropper bottles in the desk set should not be allowed to tip over.

Reagent Solutions for Anion Chemistry

Desk set: $1\,M$ HNO$_3$ KI/I$_2$ $0.1\,M$ AgNO$_3$ $0.1\,M$ BaCl$_2$
Reagent shelf: $18\,M$ H$_2$SO$_4$ (not to be put in desk set)
            Copper turnings

Each of the following is available as a solid salt and in $0.1\,M$ solution:

| NaBrO$_3$ | NaBr | Na$_2$CO$_3$ | NaCl | NaI | NaNO$_3$ |
| NaNO$_2$ | Na$_2$SO$_4$ | Na$_2$S | Na$_2$SO$_3$ | Na$_2$S$_2$O$_3$ | |

12.4 | EXPERIMENTAL TECHNIQUES

Before beginning experimental work, read through the list of techniques in Section 7.4. Proper handling of reagents and reaction mixtures is essential in avoiding errors and saving time. (It is recommended that you review Section 7.4, even if you have already completed work in Part II.)

Anion Chemistry | CHAPTER 13

13.1 GENERAL PLAN FOR ANION IDENTIFICATION

Anions participate in a variety of interesting reactions in water solution. A general survey of these reactions is presented in Chapters 1, 2, and 3. The objective of the experiments in this chapter is to carry out and describe some important examples of these reactions. Anions may be identified by the characteristic reactions which they undergo. Although it is possible to separate anions prior to identification, the procedures in this chapter do not involve time-consuming separations of anion mixtures. Sufficient information can be provided by test reactions on individual ions.

The specific problem is one of identifying a particular anion from a group of many possible anions. The 11 anions to be studied are listed alphabetically in Table 13.1. The first step will be to examine the chemistry of anions as known sodium salts. It will become apparent that two or more ions possess certain characteristics in common. For example, chloride, bromide, and iodide form insoluble silver salts. A logical approach to anion identification is to group anions together, using similarities in chemical behavior. Groups might be based on oxidation–reduction behavior, on precipitate formation with some cation, or on anion basicity.

The approach used in these experiments is based on oxidation–reduction behavior. Consequently, the results for a given anion must fall into one of three categories. The anion may be (1) an oxidizing agent, (2) a reducing agent, or (3) neither an oxidizing nor a reducing agent. The general plan is to analyze the 11 anions into five main groups, Groups A through E.

Group A: Oxidizing anions
Group B: Reducing anions
Groups C, D, E: Anions inert to oxidation–reduction (relative to the KI/I_2 system)

  Group C: Anions forming acid-insoluble silver precipitates
  Group D: Anions forming acid-soluble silver precipitates and barium precipitates
  Group E: Anions with soluble silver and barium salts

The analyses are outlined in the anion group chart in Figure 13.1. The ions are analyzed first by oxidation–reduction behavior into Group A, Group B, and Groups C, D, and E. Subsequently, Groups C, D, and E are differentiated by precipitation behavior toward silver (Ag^+) and barium (Ba^{2+}) ions. Individual members of each group are examined with respect to precipitation behavior, also.

It is not practical, nor is it necessary, to carry out all the oxidation–reduction reactions of an ion. It is sufficient to establish the behavior of the anion with respect to a standard oxidation–reduction system. The iodide–iodine system serves as a convenient standard. The chemistry of this system is reviewed in Section 13.2, in which the iodide–iodine test reactions are referred to as **Test 2 I^-/I_2**. The precipitation tests, also described in Section 13.2, are referred to as **Test 3 Ag^+/H^+** and **Test 4 Ba^{2+}/H^+**.

The objectives of the anion experiments are to

1. *Associate each known anion with one of Groups A through E.* This requires that for each ion, one or more of the following tests be carried out: Test 1 pH, Test 2 I^-/I_2, Test 3 Ag^+/H^+, Test 4 Ba^{2+}/H^+, Test 5 H_2SO_4. These tests are described on pages 153–155.
2. *Observe and describe the products of each reaction.* A balanced equation is required for each reaction. Your observations serve as the primary guide to the nature of the products to write.
3. *Identify several unknown anions in the form of their sodium salts.* The general tests will place each unknown into one of Groups A through E. A comparison of the properties with those of the known anions will allow a positive identification of each unknown.

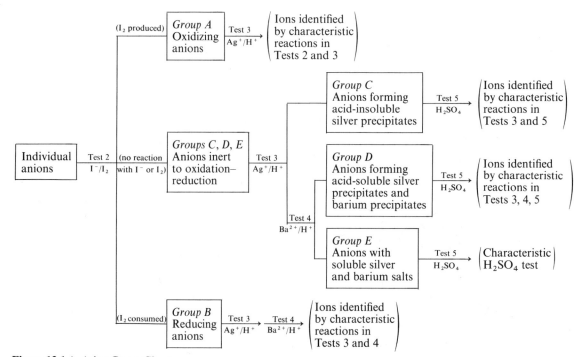

Figure 13.1 | *Anion Group Chart*

13.2 CHEMISTRY OF THE ANION TEST SYSTEMS

Test 1 pH

As discussed in Chapter 2, certain anions act as bases in water solution by reacting with H_2O molecules to release OH^-. Solutions containing these anions will give pH tests in the range 8 to 12. Anions that are very weak bases do not affect the pH of the solution, which should be essentially that of distilled water. (Na^+ is assumed to be the cation present.) The pH of distilled water is usually in the range 4 to 7, the slight acidity arising from CO_2 dissolved in the water. Basic anions will produce a blue color in litmus paper. However, the use of wide-range pH indicator paper is recommended, if available, because it provides an estimate of the actual pH range. The color scale for the wide-range paper is as follows:

pH	2	4	6	8	10
Color	red	orange	yellow	green	blue-green
	strongly acid	*acid*	*weakly acid*	*weakly basic*	*basic*

Test 2 I^-/I_2

A mixture of iodine and potassium iodide in water solution is the test reagent for oxidation–reduction. Although the element iodine occurs in compounds with a variety of oxidation numbers, the two oxidation states of importance in the anion experiments are the 0, or elemental, state and the -1 state. Elemental iodine is a slightly water-soluble solid composed of I_2 molecules. Iodine reacts readily with iodide ions to produce the triiodide ion, I_3^-:

$$I_2 + I^- \rightleftarrows I_3^- \tag{1}$$

In the presence of an excess concentration of I^- ($>0.1\ M$), this equilibrium lies largely to the right. Thus, iodine becomes soluble in aqueous solution in the form of the triiodide ion. The I_3^- ion, like I_2, is highly colored because of the absorption of light in the wavelength region below 5000 Å. Even at a concentration level of $10^{-5}\ M$, the yellow-brown color of I_3^- can easily be detected visually. By contrast, iodide ion, I^-, is colorless.

The chemistry of iodine becomes more complex in basic solution. Iodine reacts with hydroxide ion to form a mixture of $+1$ and -1 oxidation states:

$$I_2 + 2OH^- \rightarrow OI^- + I^- + H_2O \tag{2}$$

For this reason, Test 2 reactions are carried out in neutral or acid solution.

The -1 and 0 oxidation states of iodine are interrelated by the following half-reaction:

$$I_2 + 2e^- \rightleftarrows 2I^- \qquad E^0 = 0.54\ \text{volt}$$

In consulting the table of standard reduction potentials in Chapter 2, it can be seen that I_2 appears along with other oxidizing agents such as

Cl_2, O_2, and H_2O_2. Iodide ion appears along with other familiar reducing agents such as Zn, H_2, Sn^{2+}, and Fe^{2+}. Thus, the kind of reactions expected for iodine and iodide are

$$I_2 + \text{reducing agent} \rightarrow 2I^- + \text{(oxidized product)} \qquad (3)$$

$$2I^- + \text{oxidizing agent} \rightarrow I_2 + \text{(reduced product)} \qquad (4)$$

It should be easy to observe each of these reactions experimentally. As iodine is consumed by a reducing agent, its characteristic yellow-brown color disappears. Or, as iodide ion reacts with an oxidizing agent, the characteristic iodine color will appear. As indicated in the discussion of triiodide ion, both I_2 and I^- will occur together in aqueous solution. This does not alter the conclusions drawn above. A dilute mixture of I^- and I_2 will be yellow-brown. If iodide is converted to iodine, the change will be detected because the color will deepen. Indeed, if enough of the iodide is oxidized, the iodide will no longer be in excess and solid iodine will separate from the solution.

As suggested by the preceding discussion, a dilute aqueous mixture of I^- and I_2 will serve to detect the presence of both oxidizing and reducing anions. The anion to be tested is added to an I^-/I_2 solution. A deepening of the color, or precipitation of I_2, means that an oxidizing (Group A) anion is present. A decrease in intensity or complete disappearance of the color means that a reducing (Group B) anion is present. If the anion has no effect on the I^-/I_2 test solution, the anion belongs to one of the Groups C, D, E. Of course, the results obtained are relative to the I^-/I_2 system as a standard.

Test 3 Ag^+/H^+

Silver ion, Ag^+, undergoes characteristic precipitation reactions with a number of anions. The solids formed may or may not be soluble in acid solution, depending on the basicity of the anion. (See Section 2.3.) Test 3 is carried out in two steps. In the first step, a solution of $AgNO_3$ is added to the anion solution to check for precipitate formation. If a precipitate does form, then HNO_3 is added in the second step to determine the behavior of the precipitate in acid solution.

In working with Ag^+, basic solutions are avoided because Ag^+ reacts with OH^- to form a mixture of AgOH and Ag_2O. It is also important to note that Ag^+ sometimes forms complex ions as discussed in Chapters 2 and 5. In addition, Ag^+ is a reasonably good oxidizing agent:

$$Ag^+ + e^- \rightleftarrows Ag \qquad E^0 = 0.80 \text{ volt}$$

When $AgNO_3$ is added to a solution of a reducing anion, formation of elemental silver may result. In a finely divided state, silver metal appears black or dark brown.

Test 4 Ba^{2+}/H^+

Barium ion, Ba^{2+}, forms precipitates with several anions. Unlike silver ion, Ba^{2+} does not undergo reduction by anions or complex ion formation. Barium precipitates dissolve in acid solution if the anion involved is appreciably basic. Thus, Test 4 is applied in two steps. HNO_3 is added in the second step to determine the solubility of the precipitate in acid solution.

Test 5 H_2SO_4

Concentrated (18 M) sulfuric acid is a very strong proton donor. When warmed, the concentrated acid exhibits moderate oxidizing capacity as well. These properties, in addition to a high boiling point, make concentrated H_2SO_4 a useful test reagent for anion analysis. (CAUTION: *It is also a very corrosive chemical and proper care must be taken in its use.*)

H_2SO_4 converts many anions to their molecular acid form. The behavior and physical properties of the molecular acid are often characteristic of the anion. For example, addition of concentrated H_2SO_4 to NaBr produces HBr. At higher temperatures some of the HBr escapes from the mixture and may be detected by its sharp odor and its effect on pH test paper. (Moderate heating does not volatilize H_2SO_4.) Part of the HBr is oxidized by H_2SO_4. The orange coloration due to the bromine produced indicates the presence of bromide ion. The chemical changes are

$$NaBr(s) + H_2SO_4(l) \rightarrow HBr(g) + NaHSO_4(s) \quad (5)$$

$$2HBr + H_2SO_4 \rightarrow Br_2(l) + SO_2(g) + 2H_2O \quad (6)$$

The reaction of H_2SO_4 with nitrate salts serves as the basis for the specific test for the NO_3^- anion. When a nitrate salt is warmed with H_2SO_4, molecular HNO_3 is produced. The volatile HNO_3 vapor is detected by its very rapid and characteristic reaction with copper metal. The red-brown vapor of the NO_2 produced is readily identified:

$$NaNO_3(s) + H_2SO_4(l) \xrightarrow{heat} HNO_3(g) + NaHSO_4(s) \quad (7)$$

$$4HNO_3(g) + Cu(s) \longrightarrow Cu(NO_3)_2 + 2NO_2(g) + 2H_2O \quad (8)$$

13.3 ┊ PROCEDURE FOR KNOWN ANIONS

Tests 1, 2, 3, and 4 are to be carried out on 0.1 M solutions of the sodium salts of the anions. Solid sodium salts are used for Test 5. The objective is to obtain enough information on the known anions to complete Table 13.1 and to set up an anion group chart such as that in Figure 13.1.

Table 13.1

In Table 13.1, the 11 anions are identified by letters a through k. The notations that appear in the rectangular areas, for example 1a, 2b, etc., refer to Test 1 performed on anion a, Test 2 performed on anion b, etc. The blank rectangles in the table indicate which tests are to be performed. Tests corresponding to the shaded rectangles need

Anion Chemistry

Table 13.1 | *Known Anion Tests*

Anion	Test 1 pH	Test 2 I^-/I_2
a. BrO_3^- Bromate	1a	2a
b. Br^- Bromide	1b	2b
c. CO_3^{2-} Carbonate	1c	2c
d. Cl^- Chloride	1d	2d
e. I^- Iodide	1e	2e
f. NO_3^- Nitrate	1f	2f
g. NO_2^- Nitrite	1g	2g
h. SO_4^{2-} Sulfate	1h	2h
i. S^{2-} Sulfide	1i	2i
j. SO_3^{2-} Sulfite	1j	2j
k. $S_2O_3^{2-}$ Thiosulfate	1k	2k

	Test 3 Ag$^+$/H$^+$	Test 4 Ba^{2+}/H$^+$	Test 5 H$_2$SO$_4$
BrO$_3^-$	3a	4a	5a
Br$^-$	3b	4b	5b
CO$_3^{2-}$	3c	4c	5c
Cl$^-$	3d	4d	5d
I$^-$	3e	4e	5e
NO$_3^-$	3f	4f	5f
NO$_2^-$	3g	4g	5g
SO$_4^{2-}$	3h	4h	5h
S^{2-}	3i	4i	5i
SO$_3^{2-}$	3j	4j	5j
S$_2$O$_3^{2-}$	3k	4k	5k

not be carried out. Begin with Tests 1 and 2 on known solutions of the 11 anions. These solutions are available in dropper bottles in the reagent supply area. Specific directions for each test are given on page 159.

The results of each test should be recorded immediately in the corresponding space in Table 13.1. Be specific in what you record.

Precipitate formation:

Color?
Tendency to settle or remain suspended?
Coarseness?
Solubility in acid?

(Try to be precise in recording colors. For example, "yellow" colors can range from a light cream shade to bright gold. Comparative notes may also be helpful: for example, "lighter than the precipitate in Test ___.")

Gas formation:

Color?
Odor or sharpness?
Acidity? (use pH paper moistened and held above the test tube)

Figure 13.2—Known Anion Product Chart

The test results from Table 13.1 are to be organized in schematic form in Figure 13.2. The general form should be the same as that shown in Figure 13.1. However, the known anion product chart should show the specific location, chemical formula, and color of each anion product that is formed in the various tests. As an example, the results for BrO_3^- are shown in printed form in Figure 13.2. The bromate ion is put into Group A because it produces iodine in Test 2. Addition of Ag^+ to BrO_3^- in Test 3 results in the formation of a white solid, $AgBrO_3$, which does not redissolve in acid. The other anions are to be entered on one of the lined spaces within the proper group in Figure 13.2. (Extra lined spaces are provided; thus, not all of them will be used in completing the chart.)

Table 13.2

After completing the tests on the known anions, fill out Table 13.2. When completed, the equations in this table will serve as a useful summary of the chemistry involved. The notations 1a, 2b, etc., refer to the corresponding tests in Table 13.1. The steps below should be followed in completing Table 13.2.

1. Use net-ionic form for the equations. Write the correct formula for each reactant. If the reaction takes place in acid solution, the correct form of the anion depends on its basicity (Test 1). For example, if the basic sulfide ion is a reactant in acid solution, the correct formula is H_2S rather than S^{2-}.
2. Write correct formulas for the products based on your observations. Note that the reactions will be of the oxidation–reduction type in

Test 2 and, in some cases, in Test 5 also. Indicate gas evolution with the symbol (g) and precipitate formation with the symbol (s).

3. Balance the resulting equations using the techniques described in Chapter 3.

13.4 PROCEDURE FOR UNKNOWN ANIONS

The instructor will provide you with a set of unknowns. Each unknown consists of the sodium salt of a single anion. To prepare a solution of each salt, place enough sample into a 4-inch test tube to cover the rounded bottom of the tube. Add approximately 2 ml (40 drops) of distilled water and stir until the salt dissolves. Use 5-drop portions of this stock solution to carry out Tests 2, 3, and 4, using the procedure for the known anions. Use the Unknown Anion Test Sheets at the end of this chapter to record the results for each unknown.

Each unknown should be analyzed according to the known anion product chart in Figure 13.2. Tests 1 and 2 are to be carried out first. Based on the results, the anion can be assigned to Group A, Group B, or Groups C, D, or E. After applying Tests 3 and 4 as appropriate, it should be possible to make a tentative identification of the anion. Confirm the identification by carrying out Test 5. *Do not apply the nitrate test unless all anions other than* NO_3^- *have been eliminated.*

Test 1   pH

Using a stirring rod, place a drop of the anion solution on a piece of pH test paper. (The strip of test paper should be cut into several pieces.) Record the approximate pH indicated. Since water will test pH 4 to 6 due to dissolved CO_2, only tests that are more basic than pH 7 (or less than pH 4) are significant. If litmus paper is used, a blue color indicates the presence of a basic anion.

Test 2 I^-/I_2

Prepare a reference solution by mixing 5 drops of the KI/I_2 test solution and 5 drops of distilled water in a 3-inch test tube. Place 5 drops of the anion solution into a clean 3-inch test tube. Add 1 drop of $1\,M$ HNO_3, followed immediately by 5 drops of KI/I_2 test solution. Mix the solution using a stirring rod. The color of the resulting mixture should be compared with the color of the reference solution. A significantly lighter color, or a complete loss of color in the reaction mixture, indicates the presence of a reducing anion. A deeper color, or formation of solid I_2, indicates the presence of an oxidizing anion.

Test 3 Ag^+/H^+

Place 5 drops of the anion solution in a 3-inch test tube. Add 1 drop of $0.1\,M$ $AgNO_3$, stir, and observe the nature of any precipitate that forms. If a precipitate is present, add 4 drops of $1\,M$ HNO_3, with stirring, to test the solubility of the precipitate in acid. If a precipitate, which was

Anion Chemistry

Table 13.2 | Equations for Anion Reactions

(Write balanced, net-ionic equations as directed in Section 13.3.)

Test	Reaction	
2a	Bromate + iodide in acid	
3a	Bromate + silver ion	
3c	Carbonate + silver ion	
	Silver carbonate + H^+	
4c	Carbonate + barium ion	
	Barium carbonate + H^+	
3d	Chloride + silver ion	
5d	Sodium chloride(s) + H_2SO_4	
3e	Iodide + silver ion	
2g	Nitrate + iodide in acid	
4h	Sulfate + barium ion	
2i	Sulfide + iodine in acid	
3i	Sulfide + silver ion	
2j	Sulfite + iodine, in acid	
3j	Sulfite + silver ion	
	Silver sulfite + H^+	
4j	Sulfite + barium ion	
	Barium sulfite + H^+	
5j	Sodium sulfite(s) + H_2SO_4	
2k	Thiosulfate + iodine in acid	
3k	Formation of $Ag(S_2O_3)_2^{3-}$	

initially white, turns dark because of the formation of metallic silver, repeat Test 3 on a fresh sample. Add the HNO_3 as soon as possible to check the acid solubility of the precipitate before formation of silver occurs.

Test 4 Ba^{2+}/H^+

Place 5 drops of the anion solution into a 3-inch test tube. Add 1 drop of 0.1 M $BaCl_2$ and stir. If a precipitate forms, make a record of its properties and then add 4 drops of 1 M HNO_3 to check its solubility in acid. In some cases the solid will almost completely dissolve but the solution may remain cloudy. This behavior should be recorded and the precipitate considered to be acid-soluble.

Test 5 H_2SO_4

(*Wear safety glasses.*) Place a very small sample of the *solid* salt into a *dry* 3-inch test tube. The sample should only partly cover the rounded bottom of the tube. Hold the test tube in a vertical position and add 1 drop of 18 M H_2SO_4. Try to get the drop directly on the sample. If the drop runs down the sidewall, a second drop may be required. Do not add excess acid since this will obscure the results. If no immediate reaction occurs, heat the test tube by holding it in a hot-water bath. *Do not heat the tube directly in a flame.* (CAUTION: *Wear safety glasses. Do not work with H_2SO_4 above or near your face. The acid should not be allowed to get onto your skin, clothing, or the desk top. If it does, rinse thoroughly with water.*)

Note any change in color and the formation of a gas, if any. In detecting the odor of a gas, do not point the test tube at your face. Instead, use your hand to fan some of the air above the test tube toward your nose. If an irritating gas is released, rinse the test tube with water as soon as possible.

Specific Nitrate Ion Test

(*To be performed only on $NaNO_3$. When working with an unknown, this test should be carried out only after the possibility of other anions has been eliminated. Wear safety glasses.*)

Place a very small sample of *solid* $NaNO_3$ or *solid* unknown into a *dry* 3-inch test tube. The sample should only partly cover the rounded bottom of the tube. Add 1 drop of 18 M H_2SO_4 directly onto the sample, not on the sidewalls. Take several 2- to 3-inch strands of copper turnings and roll them into a ball with a diameter somewhat larger than that of the test tube. With a stirring rod, push the ball down to a position about $\frac{1}{2}$ inch above the sample but not touching the sample or H_2SO_4. Place the tube into a *boiling-water* bath. The water depth should be at least 1 inch. In 15–30 seconds, the formation of red-brown NO_2 gas will be observed if NO_3^- is present. Other ions, such as NO_2^- and I^-, may also produce a brown color. Thus, the possibility of ions other than NO_3^- must be eliminated by prior application of Tests 1 through 4.

162   Anion Chemistry

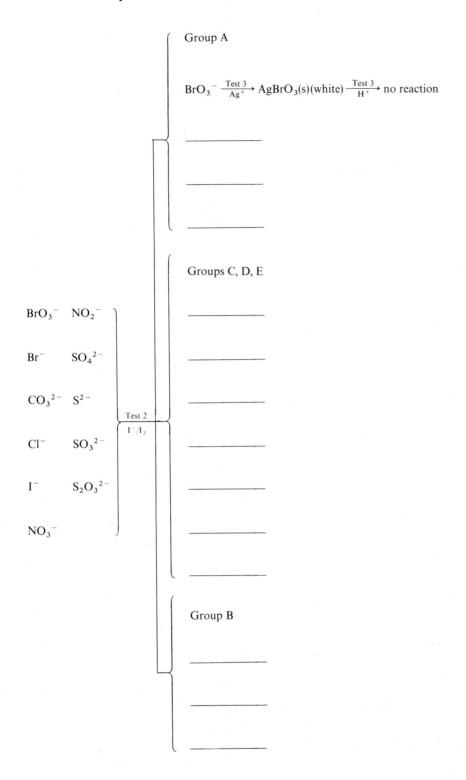

Figure 13.2 | *Known Anion Product Chart*

Figure 13.2 (*continued*) *Known Anion Product Chart*

UNKNOWN ANION TEST SHEET

NAME_____ DATE_____

ANION UNKNOWN NO._____ ANION PRESENT_____

Test 1 pH_____ Anions possible_____

Test 2 I^-/I_2_____ Group or groups possible_____

Test 3 Ag^+/H^+_____ Groups or anions possible_____

Test 4 Ba^{2+}/H^+_____ Anions possible_____
        (leave blank if not carried out)

Test 5 H_2SO_4_____ Anions possible_____

ANION UNKNOWN NO._____ ANION PRESENT_____

Test 1 pH_____ Anions possible_____

Test 2 I^-/I_2_____ Group or groups possible_____

Test 3 Ag^+/H^+_____ Groups or anions possible_____

Test 4 Ba^{2+}/H^+_____ Anions possible_____
        (leave blank if not carried out)

Test 5 H_2SO_4_____ Anions possible_____

UNKNOWN ANION TEST SHEET

NAME_____   DATE_____

ANION UNKNOWN NO._____   ANION PRESENT_____

Test 1 pH_____   Anions possible_____

Test 2 I^-/I_2_____   Groups or groups possible_____

Test 3 Ag^+/H^+_____   Groups or anions possible_____

Test 4 Ba^{2+}/H^+_____   Anions possible_____
        (leave blank if not carried out)

Test 5 H_2SO_4_____   Anions possible_____

ANION UNKNOWN NO._____   ANION PRESENT_____

Test 1 pH_____   Anions possible_____

Test 2 I^-/I_2_____   Group or groups possible_____

Test 3 Ag^+/H^+_____   Groups or anions possible_____

Test 4 Ba^{2+}/H^+_____   Anions possible_____
        (leave blank if not carried out)

Test 5 H_2SO_4_____   Anions possible_____

UNKNOWN ANION TEST SHEET

NAME_____ DATE_____

ANION UNKNOWN NO._____ ANION PRESENT_____

Test 1 pH_____ Anions possible_____

Test 2 I^-/I_2_____ Group or groups possible_____

Test 3 Ag^+/H^+_____ Groups or anions possible_____

Test 4 Ba^{2+}/H^+_____ Anions possible_____
        (leave blank if not carried out)

Test 5 H_2SO_4_____ Anions possible_____

ANION UNKNOWN NO._____ ANION PRESENT_____

Test 1 pH_____ Anions possible_____

Test 2 I^-/I_2_____ Group or groups possible_____

Test 3 Ag^+/H^+_____ Groups or anions possible_____

Test 4 Ba^{2+}/H^+_____ Anions possible_____
        (leave blank if not carried out)

Test 5 H_2SO_4_____ Anions possible_____

Problems

1. Each of four solid sodium salts is treated with concentrated H_2SO_4 at room temperature. Based on the results obtained, what anion is present in each case?

   a. A colorless, odorless gas is evolved.

   b. The solid becomes violet to brown in color and a gas with a sharp odor is produced.

   c. No gas is evolved, nor is there any change in the solid.

   d. The colorless gas evolved has a sharp odor and produces white fumes above the test tube. There is no change in the color of the solid.

2. Write the formula of one or more acids that
   a. form no common salts that are insoluble in water.

   b. form salts that react with sources of hydrogen ion to produce a colorless, odorless gas.

   c. form water-soluble barium salts and react with Pb^{2+} in acid solution to form dark-colored precipitates.

   d. are good reducing agents.

3. A series of tests are carried out in which solutions of either $AgNO_3$ or $BaCl_2$ are separately mixed with solutions of the salts indicated. Write the formula of the substance, if any, that precipitates from solution in each case.

	$AgNO_3$	$BaCl_2$
Na_2S		
$NaBrO_3$		
Na_2CO_3		
$Cu(NO_3)_2$		
H_2SO_4		
NH_4I		
$Pb(NO_3)_2$		

4. Based on the following test results, in which of Groups A to E does the phosphate, PO_4^{3-}, anion belong?

Solutions of Na_3PO_4 are very basic. When acidified, they do not react with either KI or I_2. Both Ag^+ and Ba^{2+} form water-insoluble phosphates that redissolve in acid. When Na_3PO_4 is treated with concentrated H_2SO_4, there is no apparent reaction.

Salts and Mixtures of Salts

CHAPTER 14

Using the procedures in Chapters 7 through 13 for identification of cations and anions, it should be possible to analyze a single solid salt or mixture of salts. However, the chemistry of a single salt, especially of a salt mixture, may be more complicated than that encountered in the previous chapters. Cations were studied as nitrate salts, with most of the other anions such as I^-, SO_3^{2-}, CO_3^{2-}, and NO_2^- being absent. Similarly, only the sodium salts of anions were studied in order to avoid complications arising from the cation. In analyzing salts that may contain any combination of a cation and anion, it is highly important to remember that new interactions may occur. For example, the cation present may interfere in one of the anion tests, as in the case of $AgNO_3$ as the unknown. In carrying out Anion Test 2 I^-/I_2, Ag^+ will cause precipitation of I^- as silver iodide. However, it should still be possible to make the correct identification since the precipitation of AgI has already been encountered (in Table 13.1).

Another important factor to keep in mind is the additional information that becomes available when tests are applied to an unknown salt. For example, just the fact that a salt is soluble in cold water eliminates many possible combinations of ions. If an unknown salt that is insoluble in water is found to contain CO_3^{2-}, then Na^+ and NH_4^+ ions are immediately eliminated, since their carbonates are soluble in water. The information to be gained when various reagents are being added should not be overlooked. For example, if addition of HCl to the unknown salt causes gas evolution, anions such as CO_3^{2-} may be present. The formation of brown NO_2 gas, a reduction product of HNO_3, arising from a treatment of the unknown with nitric acid, signals the presence of a reducing anion or cation. The physical appearance of the unknown will also provide information. Note the color and size of the particles. The presence of more than one kind of crystal indicates a mixture of salts.

Use the Unknown Salt Report Sheets at the end of this chapter to organize the experiments to be performed. Record all results and conclusions on these sheets. Either the cation or the anion identification may be carried out first. If results for the cation tests are inconclusive, it may be advantageous to work on the anion tests and then return to complete the cation tests.

Cation Identification

Before carrying out the cation procedures in Chapters 7 through 11, the unknown must be dissolved in a suitable solvent. Several solvents are listed below in order of preference. Do not use the more concentrated acids unless necessary. Place a small amount of solid into a 3- or 4-inch test tube, just enough to cover the rounded bottom of the test tube. Grind up the solid first if large chunks are present.

Water. Add 40 drops (approximately 2 ml) of distilled water and stir vigorously. If the solid does not dissolve in a reasonable length of time, place the test tube *in a hot-water bath* for several minutes and continue stirring. If the solid remains undissolved, try the HCl treatment below. If only part of the solid dissolves, a mixture of salts may be present. Centrifuge the mixture and save the solution for further tests. Treat the remaining solid with the solvent below. Be careful not to make the mistake of starting with so large a solid sample that the unknown appears to be insoluble in any case.

6M HCl. If the sample is not soluble in water, add 6 M HCl dropwise to the sample. The acid should be added slowly since a vigorous evolution of gas may take place. Heat in a hot-water bath if necessary. After dissolving, the mixture should be diluted with water.

6 M HNO_3. Unknowns that prove to be insoluble in HCl should be treated with 6 M HNO_3 as described above for HCl. Watch for any oxidation–reduction products that may result.

Concentrated HCl–HNO_3. Extremely insoluble salts usually can be dissolved in a mixture of hydrochloric and nitric acid, the technique used to dissolve mercury(II) sulfide in Group 2. Place a small sample into an evaporating dish and add 8 drops of 6 M HCl and 8 drops of 6 M HNO_3. (*Wear safety glasses.*) In the hood, slowly heat the mixture until it boils gently. Repeat the treatment, if necessary, by adding more 6 M HCl and 6 M HNO_3 in a 1:1 ratio. Allow the mixture to cool and then dilute it with distilled water. In some cases a new solid, such as sulfur, may be produced. Any such solid should be removed by centrifuging and kept for further tests, if necessary. (If the unknown resists treatment with acid, use the sodium carbonate method described below for anion samples.)

Anion Identification

The anion tests outlined in Chapter 13 can be employed, but keep in mind the possible influence of the cation present in these tests. Water is the preferred solvent for the anion tests. If the unknown is water-soluble, use a solution prepared as described above for the cation samples. Unknowns that are insoluble in water should be treated with sodium carbonate, as described below, to prepare a solution for the anion tests.

Na_2CO_3. Place a small quantity of the solid into an evaporating dish. The volume of the solid sample should be no more than that of 2 drops of water. Add 2 ml of water and approximately 0.5 g of solid Na_2CO_3. (*Wear safety glasses.*) Boil the mixture gently for 3 minutes, adding more water if necessary. Allow the evaporating dish to cool and then transfer

the contents to a test tube. Centrifuge. Pour off the liquid layer, which, after dilution with water, can be used for the anion tests. The remaining solid consists of a metal carbonate, or a mixture of carbonates, and possibly some undissolved unknown sample. If desired, this solid may be used to prepare a solution for cation tests. Slowly add 6 M HNO_3 to the solid until CO_2 is no longer evolved. Heat the test tube in a water bath. Dilute the mixture with water. If any solid remains, it should be removed by centrifuging and discarded.

UNKNOWN SALT REPORT SHEET

NAME_____ DATE_____

UNKNOWN SALT REPORT SHEET

NAME_____ DATE_____

UNKNOWN SALT REPORT SHEET

NAME_____ DATE_____

UNKNOWN SALT REPORT SHEET

NAME_____ DATE_____

UNKNOWN SALT REPORT SHEET

NAME_____   DATE_____

Appendix*

Dissociation Constants of Acids in Water

Acid		K_a (25°C)
Acetic	$HC_2H_3O_2$	1.8×10^{-5}
Aluminum(III), hydrated	$Al(H_2O)_6^{3+}$	1×10^{-5}
Arsenic	H_3AsO_4	6×10^{-3} (K_1)
	$H_2AsO_4^-$	2×10^{-7} (K_2)
Chromic	H_2CrO_4	Large (K_1)
	$HCrO_4^-$	3.2×10^{-7} (K_2)
Carbonic	CO_2 (or H_2CO_3)	4.3×10^{-7} (K_1)
	HCO_3^-	5×10^{-11} (K_2)
Hydrocyanic	HCN	4.9×10^{-10}
Hydrofluoric	HF	4×10^{-4}
Hydrogen peroxide	H_2O_2	2.4×10^{-12}
Hydrogen sulfide	H_2S	1×10^{-7} (K_1)
	HS^-	1×10^{-14} (K_2)
Hypochlorous	$HClO$	3.0×10^{-8}
Nitrous	HNO_2	4.6×10^{-4}
Oxalic	$H_2C_2O_4$	6×10^{-2} (K_1)
	$HC_2O_4^-$	6×10^{-5} (K_2)
Phosphoric	H_3PO_4	7.5×10^{-3} (K_1)
	$H_2PO_4^-$	6.2×10^{-8} (K_2)
Sulfuric	H_2SO_4	Large (K_1)
	HSO_4^-	1.2×10^{-2} (K_2)
Sulfurous	SO_2 (or H_2SO_3)	1.5×10^{-2} (K_1)
	HSO_3^-	1×10^{-7} (K_2)

Dissociation Constants of Bases in Water

Base		K_b (25°C)
Ammonia	NH_3	1.8×10^{-5}
Ethylenediamine	$NH_2CH_2CH_2NH_2$	5×10^{-4} (K_1)
		4×10^{-7} (K_2)
Hydroxylamine	NH_2OH	1.1×10^{-8}

* Equilibrium constants in this appendix were taken, in part, from the 48th edition of the *Handbook of Chemistry and Physics*, Chemical Rubber Publishing Company, Cleveland, 1968.

Solubility-Product Constants

Compound		K_{sp} (25°C)
Barium carbonate	$BaCO_3$	8×10^{-9}
Barium chromate	$BaCrO_4$	2×10^{-10}
Barium oxalate	BaC_2O_4	2×10^{-7}
Barium sulfate	$BaSO_4$	1.1×10^{-10}
Cadmium hydroxide	$Cd(OH)_2$	3×10^{-14}
Cadmium sulfide	CdS	8×10^{-27}
Calcium carbonate	$CaCO_3$	1×10^{-8}
Calcium chromate	$CaCrO_4$	7×10^{-4}
Calcium fluoride	CaF_2	4×10^{-11}
Calcium oxalate	CaC_2O_4	3×10^{-9}
Calcium sulfate	$CaSO_4$	2×10^{-4}
Chromium hydroxide	$Cr(OH)_3$	7×10^{-31}
Copper(II) sulfide	CuS	8×10^{-45}
Iron(II) hydroxide	$Fe(OH)_2$	2×10^{-14}
Iron(II) sulfide	FeS	4×10^{-19}
Iron(III) hydroxide	$Fe(OH)_3$	1×10^{-36}
Lead chloride	$PbCl_2$	1.6×10^{-5}
Lead chromate	$PbCrO_4$	2×10^{-14}
Lead hydroxide	$Pb(OH)_2$	3×10^{-15}
Lead sulfate	$PbSO_4$	1×10^{-8}
Lead sulfide	PbS	8×10^{-28}
Mercury(I) chloride	Hg_2Cl_2	2×10^{-18}
Mercury(I) sulfide	Hg_2S	6×10^{-44}
Mercury(II) sulfide	HgS	1×10^{-52}
Nickel hydroxide	$Ni(OH)_2$	2×10^{-16}
Nickel sulfide	NiS	2×10^{-21}
Silver bromide	$AgBr$	7.7×10^{-13}
Silver chloride	$AgCl$	1.6×10^{-10}
Silver chromate	Ag_2CrO_4	9×10^{-12}
Silver iodide	AgI	1.5×10^{-16}
Silver sulfide	Ag_2S	1×10^{-51}

RECOMMENDED EQUIPMENT AND SUPPLY ITEMS

Optional items are in parentheses.

- 10 3-inch test tubes
- 2 4-inch test tubes
- 1 50-ml beaker
- 1 100-ml (or 150-ml) beaker
- 1 (600-ml beaker)
- 1 desk set of 10 $\frac{1}{2}$-oz dropper bottles with tray or rack
- 1 10-ml graduated cylinder
- 1 porcelain evaporating dish, 75 mm
- 1 watch glass, 75-mm, to cover evaporating dish

- 1 test-tube rack for 3- and 4-inch test tubes
- 1 small test-tube holder
- 1 3-inch test-tube brush
- 1 (aluminum water-bath cover, punched to hold 3-inch test tubes)

- 6 4-inch stirring rods (or 2 ft of 3-mm soft glass rod)
- 4 pipet droppers (or 2 ft of 6-mm soft glass tubing, plus 4 dropper bulbs)
- 1 110°C thermometer
- 1 6-inch piece of stainless-steel wire
- 4 corks (or rubber stoppers) for 3-inch test tubes
- 1 vial of wide-range pH indicator paper (or neutral litmus paper)
- 1 vial of methyl violet test paper

- 1 Bunsen burner
- 1 tripod support (or a ringstand, plus 3-inch ring clamp)
- 1 4-inch-square wire gauze
- 1 4-inch-square asbestos pad
- 1 triangular file

1 (burner wing top)
1 tongs, for evaporating dish
1 stainless-steel or nickel spatula
1 (250-ml, or 500-ml, polyethylene wash bottle)

2 towels
matches, labels, (pipe cleaners)
safety glasses
analytical centrifuges—approximately 1 for every 6 people

REAGENTS FOR CATION CHEMISTRY—PART II

Primary Solutions

All solutions are to be made using distilled water, except as indicated. Solutions marked with an asterisk (*) deteriorate slowly; they should be prepared shortly before use.

Acetic acid (6 M)	$HC_2H_3O_2$	Dilute 345 ml of glacial acetic acid (17 M) to 1 liter
Ammonia (2 M)	NH_3	Dilute 132 ml of 28% ammonia (15 M) to 1 liter
Ammonia (6 M)	NH_3	Dilute 400 ml of 28% ammonia (15 M) to 1 liter
Ammonium chloride (3 M)	NH_4Cl	160 g, dilute to 1 liter
Hydrochloric acid (2 M)	HCl	Dilute 167 ml of concentrated (12 M) hydrochloric acid to 1 liter
Hydrochloric acid (6 M)	HCl	Dilute 500 ml of concentrated (12 M) hydrochloric acid to 1 liter
Nitric acid (1 M)	HNO_3	Dilute 63 ml of concentrated (16 M) nitric acid to 1 liter
Nitric acid (6 M)	HNO_3	Dilute 375 ml of concentrated (16 M) nitric acid to 1 liter
Potassium chromate (0.1 M)	K_2CrO_4	19 g, dilute to 1 liter
*Thioacetamide (8%)	$CH_3C(S)NH_2$	80 g, dilute to 1 liter

Secondary Reagents (*Amber bottles*)

The order of these reagents corresponds to that recommended on page 73. All solutions are to be made using distilled water, except as indicated. Solutions marked with an asterisk (*) deteriorate slowly; they should be prepared shortly before use.

		Grams/Liter of Solution
Silver nitrate (0.1 M)	$AgNO_3$	17
Mercurous nitrate (0.03 M)	$Hg_2(NO_3)_2 \cdot 2H_2O$	17

Add the $Hg_2(NO_3)_2 \cdot 2H_2O$ to 0.6 M HNO_3 and stir and heat until the solid dissolves. (To prepare 0.6 M HNO_3, add 37 ml of 16 M HNO_3 to 500 ml of water and dilute with water to 1 liter.)

Lead nitrate (0.1 M)	$Pb(NO_3)_2$	33
Arsenic(III) (0.05 M) (arsenious oxide)	As_2O_3	5

Add the As_2O_3 to 500 ml of 3 M HCl. Stir and heat until the solid dissolves, then dilute with water to 1 liter.

Mercuric nitrate (0.1 M)	$Hg(NO_3)_2 \cdot H_2O$	34

Dissolve the $Hg(NO_3)_2 \cdot H_2O$ in 0.1 M HNO_3. (To prepare 0.1 M HNO_3, add 6 ml of 16 M HNO_3 to 500 ml of water and dilute with water to 1 liter.)

Cupric nitrate (0.1 M)	$Cu(NO_3)_2 \cdot 3H_2O$	24
Cadmium nitrate (0.1 M)	$Cd(NO_3)_2 \cdot 4H_2O$	31
Ferric nitrate (0.1 M)	$Fe(NO_3)_3 \cdot 9H_2O$	40

Dissolve the $Fe(NO_3)_3 \cdot 9H_2O$ in 0.1 M HNO_3.

		Grams/Liter of Solution
Nickel nitrate (0.1 M)	Ni(NO$_3$)$_2$·6H$_2$O	29
Aluminum nitrate (0.2 M)	Al(NO$_3$)$_3$·9H$_2$O	75

Dissolve the Al(NO$_3$)$_3$·9H$_2$O in 0.1 M HNO$_3$.

Chromic nitrate (0.1 M)	Cr(NO$_3$)$_3$·9H$_2$O	40

Dissolve the Cr(NO$_3$)$_3$·9H$_2$O in 0.1 M HNO$_3$.

Barium nitrate (0.1 M)	Ba(NO$_3$)$_2$	26
Calcium nitrate (0.2 M)	Ca(NO$_3$)$_2$·4H$_2$O	48
Ammonium nitrate (0.3 M)	NH$_4$NO$_3$	24
Sodium nitrate (0.1 M)	NaNO$_3$	9
Hydrochloric acid (12 M)	HCl	—
Sulfuric acid (18 M)	H$_2$SO$_4$	—
Sodium hydroxide (6 M)	NaOH	240
Ammonium acetate (2 M)	NH$_4$C$_2$H$_3$O$_2$	154
Ammonium carbonate (2 M)	(NH$_4$)$_2$CO$_3$·H$_2$O	228

Dissolve the (NH$_4$)$_2$CO$_3$ in a mixture of 80 ml of 15 M NH$_3$ and 400 ml of water and then dilute with water to 1 liter.

Ammonium oxalate (0.5 M)	(NH$_4$)$_2$C$_2$O$_4$·H$_2$O	71
Ammonium thiocyanate (0.1 M)	NH$_4$SCN	8
Lead acetate (0.1 M)	Pb(C$_2$H$_3$O$_2$)$_2$·3H$_2$O	38
*Potassium ferrocyanide (0.05 M)	K$_4$Fe(CN)$_6$·3H$_2$O	21
*Stannous chloride (0.2 M)	SnCl$_2$·2H$_2$O	45

Dissolve the SnCl$_2$·2H$_2$O in a mixture of 170 ml of 12 M HCl and 400 ml of water and then dilute with water to 1 liter. Add enough granulated tin metal to partially cover the bottom of the container. Tin metal should also be added to the individual dropper bottles on the reagent shelf.

Aluminon (0.1%)	C$_{22}$H$_{23}$N$_3$O$_9$	1

(ammonium salt of aurintricarboxylic acid)

Dimethylglyoxime (1%)	(CH$_3$)$_2$C$_2$(NOH)$_2$	10

Dissolve this reagent in 95% ethanol.

*Hydrogen peroxide (3%)	H$_2$O$_2$	—
Manganese dioxide (solid)	MnO$_2$	—
Sodium dithionite (solid)	Na$_2$S$_2$O$_4$	—

Group Known Solutions

Group 1: Ag^+, Hg_2^{2+}, Pb^{2+}

		Grams/Liter of Solution
Silver nitrate (0.10 M)	AgNO$_3$	17
Mercurous nitrate (0.03 M)	Hg$_2$(NO$_3$)$_2$·2H$_2$O	17
Lead nitrate (0.10 M)	Pb(NO$_3$)$_2$	33

Dissolve in 0.6 M HNO$_3$. (To prepare 0.6 M HNO$_3$, add 37 ml of 16 M HNO$_3$ to 500 ml of water and dilute with water to 1 liter.) Dissolve the mercurous nitrate first, stirring and, if necessary, heating until the salt dissolves. Dissolve each of the other salts, in turn, and mix thoroughly.

Group 2: Hg^{2+}, Pb^{2+}, Cu^{2+}, Cd^{2+} [As(III) to be added separately]

Mercuric nitrate (0.10 M)	Hg(NO$_3$)$_2$·H$_2$O	34
Lead nitrate (0.10 M)	Pb(NO$_3$)$_2$	33
Cupric nitrate (0.10 M)	Cu(NO$_3$)$_2$·3H$_2$O	24
Cadmium nitrate (0.10 M)	Cd(NO$_3$)$_2$·4H$_2$O	31

Dissolve in 0.1 M HNO$_3$. (To prepare 0.1 M HNO$_3$, add 6 ml of 16 M HNO$_3$ to 500 ml of water and dilute with water to 1 liter.) Dissolve each salt, in turn, and mix thoroughly. [Arsenic (III) is to be added separately, as specified in the procedure for the Group 2 known].

Group 3: Fe^{3+}, Ni^{2+}, Al^{3+}, Cr^{3+}

Ferric nitrate (0.10 M)	$Fe(NO_3)_3 \cdot 9H_2O$	40
Nickel nitrate (0.10 M)	$Ni(NO_3)_2 \cdot 6H_2O$	29
Aluminum nitrate (0.20 M)	$Al(NO_3)_3 \cdot 9H_2O$	75
Chromic nitrate (0.10 M)	$Cr(NO_3)_3 \cdot 9H_2O$	40

Dissolve in 0.1 M HNO_3, prepared as above.

Groups 4 and 5: Ba^{2+}, Ca^{2+}, NH_4^+, Na^+

Barium nitrate (0.10 M)	$Ba(NO_3)_2$	26
Calcium nitrate (0.20 M)	$Ca(NO_3)_2 \cdot 4H_2O$	48
Ammonium nitrate (0.20 M)	NH_4NO_3	16
Sodium nitrate (0.10 M)	$NaNO_3$	9

Dissolve in water.

REAGENTS FOR ANION CHEMISTRY—PART III

Solutions for the Desk Set

All solutions are to be made using distilled water, except as indicated. Solutions marked with an asterisk (*) deteriorate slowly; they should be prepared shortly before use.

Grams/Liter of Solution

Nitric acid (1 M) HNO_3

Add 63 ml of 16 M HNO_3 to 500 ml of water and then dilute with water to 1 liter.

*Potassium iodide (0.2 M)/iodine (0.01 M) reagent KI/I_2 KI 33 ; I_2 2.5

Add the KI and I_2 to approximately 50 ml of water, stir until all the solids dissolve, and then dilute with water to 1 liter.

Silver nitrate (0.1 M) (amber bottle)	$AgNO_3$	17
Barium chloride (0.1 M)	$BaCl_2 \cdot 2H_2O$	24

Solutions for the Reagent Shelf (*Amber bottles*)

All solutions are to be made using distilled water, except as indicated. Solutions marked with an asterisk (*) deteriorate slowly; they should be prepared shortly before use.

Grams/Liter of Solution

Sulfuric acid (18 M)	H_2SO_4	—
Sodium bromate (0.1 M)	$NaBrO_3$	15
Sodium bromide (0.1 M)	$NaBr$	10
Sodium carbonate (0.1 M)	Na_2CO_3	11
	(or $Na_2CO_3 \cdot H_2O$)	(12)
Sodium chloride (0.1 M)	$NaCl$	6
Sodium iodide (0.1 M)	NaI	15
Sodium nitrate (0.1 M)	$NaNO_3$	9
Sodium nitrite (0.1 M)	$NaNO_2$	7
Sodium sulfate (0.1 M)	Na_2SO_4	14
	(or $Na_2SO_4 \cdot 10H_2O$)	(32)
*Sodium sulfide (0.1 M)	$Na_2S \cdot 9H_2O$	24
*Sodium sulfite (0.1 M)	Na_2SO_3	13

This solution is susceptible to air oxidation. It should be prepared immediately before use. Alternatively, students can prepare this solution by dissolving 1.3 g of Na_2SO_3 in 100 ml of water.

Sodium thiosulfate (0.1 M)	$Na_2S_2O_3 \cdot 5H_2O$	25

Small bottles of the following solid reagents should be available in the reagent area.

Copper turnings	Cu		
Sodium bromate	$NaBrO_3$	Sodium nitrite	$NaNO_2$
Sodium bromide	NaBr	Sodium sulfate	Na_2SO_4
Sodium carbonate	Na_2CO_3	Sodium sulfide	$Na_2S \cdot 9H_2O$
Sodium chloride	NaCl	Sodium sulfite	Na_2SO_3
Sodium iodide	NaI	Sodium thiosulfate	$Na_2S_2O_3$
Sodium nitrate	$NaNO_3$		

Index

Abbreviations, (s), (g), (aq), 13
Absorption of light, 62
Acetate hydrolysis, 11
Acetic acid, 8–9, 11, 23–24, 32–35, 100, 133
Acidity, 36, 75, 153
Acids
    cations as acids, 10
    definitions, 8–9
    dissociation, 10, 32–35
    effect on solubility of salts, 13
    naming, 4–5
$Ag(NH_3)_2^+$, 53, 54, 64, 79–80
$Ag(S_2O_3)_2^{3-}$, 18
$Al(OH)_4^-$, 111, 116
Aluminon, 111, 117
Aluminum ion, 10, 111
Aluminum oxide, hydrated, 111
Amidochloride complex of mercury(II), 80–81
Ammine ligands, 51
Ammonia, 9, 34, 64, 129
    dissociation, 34
Ammonium chloride, 102, 117, 131
Ammonium ion, 129, 131
Amphoteric properties, 111
Anion group chart, 152
Anions
    base strength, 12
    definition, 2
    formulas, 4
    as ligands, 18, 50
    naming, 3
    precipitation, 12
    reaction as bases, 10
    reaction as oxidizing or reducing agents, 14
Aquo ligands, 51
Arsenic(III), 93
AsS_3^{3-}, 93

Balancing equations, 23–28
Barium, 127–128, 154–155
Bases
    anions as bases, 10
    definitions, 8–9
    dissociation, 10, 32–35

Bauxite ore, 111
Bidentate ligand, 51
Bimolecular process, 66
Bonding in complex ions, 59–62
Bromate reactions, 158, 162
Brønsted–Lowry, definition of acids and bases, 8
Buffer solutions, 34–35, 109

Cadmium, 95
Calcium, 127–128
Calculations
    buffer solution, 34–35
    dissociation of weak acids and bases, 32–33
    extent of precipitation, 39–41
    pH, 35–36
    ratio of $[Ag^+]$ to $[Ag(NH_3)_2^+]$ in NH_3, 64
    solubility, 37–39
Carbon monoxide, 60
Carbonate ion, reaction with Ba^{2+} and Ca^{2+}, 127
Cations
    complex-ion formation, 18
    in coordination compounds, 49
    definition, 1
    formulas, 4
    naming, 3
    precipitation, 12
    reaction as acids, 10
    reaction as oxidizing or reducing agents, 14
Centrifuge, 75
Chelates, 50
Chloro complexes, in HCl, 83
Chromate ion, 112
    reaction with Ba^{2+} and Ca^{2+}, 127–128
    reaction with Pb^{2+}, 81, 112
Chromium, 111–113
Cinnabar, 92
Cis-trans isomers, 55–56
$Co(CN)_5H_2O^{2-}$, 65
Color
    of complex ions, 62–63
    of precipitates, 158
    of silver salts, 80

of sulfides, 91
Common-ion effect, 39
Complex ions, 18, 49–50, 52–54
    bonding, 60–62
    charge of, 53
    colors, 62–63
    coordination number, 52
    dissociation, 64
    electron configurations, 63–64
    magnetic properties, 63–64
    naming, 51
    oxidation–reduction, 66
    substitution reactions, 64–66
Concentrations
    calculation, 32
    change, 8
$Co(NH_3)_5Cl^{2+}$, 65
Conjugate acids and bases, 9, 11
Coordination compounds, 49–51
    isomerism, 54–56
Coordination numbers, 52–53
Copper, 53–54, 94–95
Copper metal, formation by dithionite, 95
Copper turnings, in nitrate test, 161
$Cr(C_2O_4)_3^{3-}$, 56
$Cr(H_2O)_6^{3+}$, 62, 111
$Cr(H_2O)_6Cl_3$, 55
$Cr(H_2O)_4Cl_2^+$, 55–56
$Cr(H_2O)_2(OH)_4^-$, 67, 112, 126
$Cr_2O_7^{2-}$, 112–113
Crystal field theory, 60–64
$Cu(NH_3)_4(H_2O)_2^{2+}$, 53, 94–95, 100
Cyanide ion, 60

d orbitals, 59–64
Diamagnetic ions, 63–64, 67
Dichromate ion, 112–113
Dimethylglyoxime, 53–54, 111, 116
Dissociation
    buffer solution, 34
    constants, *table*, 187
    percent, 33
    repression of in buffers, 35
Distilled water, 74, 189
Donor atom, 50
Dq, 62
Droppers, 71–72

e_g orbitals, 62
Electrode potentials, 16
Electron
    configuration, 59, 63–64
    d orbitals, 60–61
    energy levels, 61–62
    pairing, 63
Electron-transfer reactions, 66
Equations
    half-reactions, 27–28

    net-ionic, 23
    nonredox, 23
    oxidation–reduction, 24–28
Equilibrium, 7–8, 31
    acid–base, 8–10, 32–35
    calculations, 32
    dynamic, 7, 12
    solubility, 12, 37–39
Equilibrium-constant expression, 7, 31
Equilibrium shifts, 8
Equipment, 188-189
Ethylenediamine, 50, 55, 57
Evaporating solutions, 75–76
Excitation of electrons, 62–64
Exponentials, 35–36
Eye protection, 73

$Fe(H_2O)_5NCS^{2+}$, 18, 51, 65, 68, 110, 116
Ferric ion [*see* iron(III)]
Ferrocyanide ion, 51, 66, 100
Flame tests, 128–129, 132–134
Formulas of ions, 3–5

Galvanic cells, 16–17
General cation diagram, 2
General unknown, 2, 84, 139–143
Geometrical isomers, 55–58
Geometries of complex ions, 52–54
Group diagrams
    Group 1, 82
    Group 2, 96
    Group 3, 114
    Groups 4 and 5, 130
Group known solutions, preparation, 190–191
Group separation
    Group 1, 79
    Group 2, 91
    Group 3, 109
    Groups 4 and 5, 127

Half-reactions, 15–18, 27–28
HCl–HNO_3 (aqua regia), 92, 174
Heating test tubes, 73
Hg_2^{2+} cation, properties, 80
$HgCl_4^{2-}$, 92
$Hg(NH_2)Cl$, 80
HgS_2^{2-}, 92
High-spin configuration, 63
Hot-water bath, 74
Hydrate isomers, 54–55
Hydrochloric acid
    cleaning flame-test wire, 132–133
    Group 1 precipitation, 79-81, 83
    solvent for unknown salts, 174
Hydrogen peroxide, 112–113, 115
Hydrogen sulfide, 17, 41–43, 91–92